Recent Trends in Plant Protection

NIPA® GENX ELECTRONIC RESOURCES & SOLUTIONS P. LTD.
New Delhi-110 034

About the Editors

Dr. Mukesh Kumar Mishra is working as Assistant Professor in Department of Entomology, College of Agriculture, Banda University of Agriculture & Technology, Banda (Uttar Pradesh). He did his post-graduation and Ph.D. in Agricultural Entomology from A.N.D.U.A. & T., Kumarganj, Ayodhya (U.P.). He has qualified ICAR-NET and received different awards from different professional societies. He has over 12 years of experience in the field of Entomology in various capacities. He has published 28 research papers in the reputed scientific journals, 08 popular articles, 2 book chapters and 1 book.

Mr. Arun Kumar is a research scholar at Department of Entomology, CSAUA&T, Kanpur-208002 (U.P) since 2021. He completed his graduation from Institute of Agricultural Sciences, B.U., Jhansi in 2019. Thereafter, he completed his post-graduation at BUA&T, Banda-210001, Uttar Pradesh in 2021 and qualified ASRB NET in 2021 & 2023. He has 58 Popular Articles, 6 Review Article, 6 Abstracts, 7 Chapters, 2 Edited Books and one Research Paper in various National and International Magazines/Journals/Souvenir etc.

Mr. Saurabh Kumar is a field development officer at research and development department, FMC Private Ltd. He did his graduation and Post graduation (Entomology) From Banda University of Agriculture & Technology, Banda, Uttar Pradesh (210001), India. He has 15 Popular Articles, 1 Abstract, 1 Chapter and 2 Research Papers.

Recent Trends in Plant Protection

Mukesh Kumar Mishra
Arun Kumar
Saurabh Kumar

NIPA® GENX ELECTRONIC RESOURCES & SOLUTIONS P. LTD.
New Delhi-110 034

NIPA® GENX ELECTRONIC RESOURCES & SOLUTIONS P. LTD.

101,103, Vikas Surya Plaza, CU Block
L.S.C.Market, Pitam Pura, New Delhi-110 034
Ph : +91 11 27341616, 27341717, 27341718
E-mail:newindiapublishingagency@gmail.com
www: www.nipabooks.com

For customer assistance, please contact
Phone: + 91-11-27 34 17 17
Fax: + 91-11-27 34 16 16
E-Mail: feedbacks@nipabooks.com

Print ISBN: 978-93-58879-78-0

ebook ISBN: 978-93-58879-04-9

Composed and Designed by NIPA.

Preface

The book "**Recent Trends in Plant Protection**" is written with the comprehensive information on modern and highly profitable techniques for pest management including insect- pest and diseases. The content of this book are presented in simple language and easy to understand, which not only arouses students' interest but also simplifies the comprehension process. The chapters of this book provides an overview of the development of modern pest management approaches *viz*., plant diseases detection techniques, insect trapping, bio inputs and bio pesticides (importance of predatory insects including its conservation techniques, important entomophagous pathogens and their mass multiplication), impact of agrochemicals, new classifications of pesticides and pesticide spraying equipment, plant vectors, regulatory and molecular approaches and non-insect-pest management. A chapter of this book also being discussed to plant and natural enemies' relationship. A part of this topic also covers the chapters on edible and commercial importance insects like honey bee, silk worm and lac insect. It will be great help to students and teachers in the field of agriculture particularly Crop Protection throughout. The purpose of this book is to bridge the gap between research and practice, showcasing innovative solutions and promoting interdisciplinary collaboration. It provides an avenue for sharing experiences, discussing challenges, and exploring new avenues in plant protection research and implementation. We would like to express our gratitude to all the authors who contributed their expertise and valuable insights to this book. Their dedication and commitment to advancing the field of plant protection have made this compilation possible. We hope that this book will serve as a valuable resource and inspire further research and innovations in the field of plant protection. Together, we can work towards sustainable agriculture, enhanced crop protection, and a more food-secure future.

Editors

Contents

1

Advanced Techniques for Plant Disease Detection

Deep Narayan Mishra[1*] and Ira Sharma[2]

[1]Division of Plant Pathology, ICAR-Indian Agricultural Research Institute New Delhi-110012, India*

[2]Department of Plant Breeding and Genetics, CSKHPKV, Palampur-176062 Himachal Pradesh, India

Introduction

Plant disease detection is a crucial aspect of modern agriculture, as it allows early identification and management of diseases, minimizing crop losses and maximizing yields. In recent years, advancements in technology have led to the development of more sophisticated techniques for plant disease detection. Point-of-care (POC) diagnostic assay for plant disease detection is a rapid, advanced, and on-site testing method designed to identify the presence of pathogens or diseases in plants. This type of diagnostic tool is particularly valuable for farmers, agronomists, and researchers who need immediate results to make timely decisions about disease management and crop protection strategies. POC assays aim to provide accurate and reliable results quickly, without the need for specialized equipment or laboratory facilities. POC diagnostic assay can be performed swiftly and cheaply in the field as it does not require sophisticated lab equipment (Yager *et al.*, 2006). PCR-based techniques are superior to other technologies in many ways, but they are significantly less suitable for POC applications because they need power for performing the temperature alterations necessary for DNA amplification. Isothermal DNA amplification techniques are a suitable alternative to overcome this constraint (Gill and Ghaemi, 2008; Craw and Balachandran, 2012). For POC detection of pathogen DNA, isothermal amplification in conjunction with lateral flow strips and portable fluorometers has been utilised successfully (Notomi *et al.*, 2000; Vincent *et al.*, 2004; Piepenburg *et al.*, 2006; Chow *et al.*, 2008; Lutz *et al.*, 2010; Mahalanabis *et al.*, 2010; Rohrman and Richards-Kortum, 2012). Fluorometers are quite expensive and are not best suited for use in fields having adverse weather conditions. Nucleic acid-based techniques have revolutionized the field of plant disease detection due to their high specificity and sensitivity. These techniques rely on the detection and analysis of specific nucleic acids, such as DNA and RNA, to identify pathogens responsible for plant diseases. These nucleic acid-based techniques provide accurate and reliable results, enabling early detection of plant diseases, prompt management decisions, and preventing potential crop losses due to pathogen infections. They have become indispensable tools for plant pathologists and plant health management programs worldwide. Some of the most used nucleic acid-based techniques for plant disease detection include:

Polymerase Chain Reaction (PCR): PCR is a well-established technique that amplifies specific DNA sequences. It requires primers designed to target unique regions of the pathogen's genome. PCR can detect pathogens even at low concentrations and is widely used for various plant pathogens. The PCR reaction is set up, which includes the extracted DNA/RNA, the designed primers, a DNA polymerase enzyme, and a mixture of nucleotides (dNTPs). The PCR machine goes through cycles of heating and cooling to facilitate the amplification of the target DNA or RNA sequence.

- Denaturation: The reaction mixture is heated to a high temperature (around 94-98°C), which causes the double-stranded DNA to separate into single strands.
- Annealing: The reaction is then cooled to a lower temperature (around 50-65°C), allowing the primers to bind to their complementary sequences on the target DNA or RNA.
- Extension: The reaction is heated to a moderate temperature (around 72°C), and the DNA polymerase synthesizes a new DNA strand that complements the target DNA or RNA, starting from the primers.
- Amplification Detection: The amplification of the target sequence occurs exponentially during the PCR cycles. If the target pathogen's DNA or RNA is present in the sample, the amount of the amplified product increases significantly with each cycle.
- Gel Electrophoresis: After the PCR is completed, the amplified products are separated and visualized using gel electrophoresis. The PCR products are loaded into a gel, and an electric field is applied to separate the DNA fragments based on their size. A DNA ladder is also run on the gel to determine the size of the amplified products. If the pathogen's DNA or RNA was present in the sample, the corresponding amplified band will be visible on the gel.
- Result Interpretation: The presence or absence of the specific amplified band indicates the presence or absence of the target pathogen in the plant sample.

Reverse Transcription PCR (RT-PCR): RT-PCR is used to detect RNA-based pathogens (e.g., RNA viruses) by first converting RNA into complementary DNA (cDNA) using reverse transcriptase enzyme. The resulting cDNA is then amplified using PCR.

Real-Time PCR (qPCR): Real-time PCR is a variation of PCR that allows quantification of the pathogen's DNA or RNA during the amplification process. It provides information about the initial amount of pathogen present in the sample, making it useful for quantification and monitoring disease progression. Steps of qPCR are as follows:

- Sample Collection and Preparation: The first step is to collect plant samples suspected of being infected by a pathogen. These samples could be leaves, stems, fruits, or any other plant part showing disease symptoms. Proper sample preparation is essential to ensure the quality of extracted genetic material.
- RNA Extraction: qRT-PCR detects RNA, so the genetic material needs to be extracted from the collected plant samples. RNA extraction methods may vary depending on the specific plant species and pathogen involved. Commercial RNA extraction kits are commonly used for this purpose.
- cDNA Synthesis: The extracted RNA is then reverse transcribed into complementary DNA (cDNA) using a reverse transcriptase enzyme. This step converts the RNA template into DNA, which will serve as a template for the subsequent PCR amplification.
- Primer and Probe Design: Specific primers and a probe (also known as a TaqMan probe) are designed based on the unique genetic sequences of the target pathogen. These primers and the probe are essential for selectively amplifying and detecting the pathogen's DNA during the qRT-PCR process.
- qRT-PCR Setup: The qRT-PCR reaction mixture is prepared, containing the cDNA template, primers, probe, and the qPCR master mix, which includes the DNA polymerase and fluorescent dyes. The reaction mixture is dispensed into individual wells of a qRT-PCR plate.
- Amplification and Detection: The qRT-PCR instrument cycles through a series of temperature changes, repeatedly amplifying the target DNA and allowing the fluorescence signal to be detected in real-time. The instrument measures the fluorescence emitted by the dye-bound probes at each cycle, and this fluorescence is proportional to the amount of amplified DNA. The data collected during this step is used to generate a quantification cycle (Ct) value.
- Data Analysis: The Ct values obtained during qRT-PCR are used to determine the quantity of the target pathogen in the original sample.

Comparing the Ct values to a standard curve generated from known concentrations of the target pathogen can help estimate the pathogen load in the sample.

Loop-mediated isothermal amplification (LAMP)

LAMP is a molecular method for nucleic acid amplification that is highly selective as it utilises an array of four different primers (forward and reverse and external and internal) that bind to six (or eight) distinct regions on the target gene (Notomi *et al.*, 2000). Under isothermal conditions, LAMP reactions can be carried out easily by utilising Bst DNA polymerase which is known to have high-displacement activity (Chen *et al.*, 2015; Notomi *et al.*, 2000). Unlike PCR, LAMP can be performed at a steady temperature, with a short reaction time. Amplification can be visually detected or quantified using the turbidity or colour of the reaction mixture (Mori *et al.*, 2001, 2004; Iwamoto *et al.* 2003; Goto *et al.*, 2009).

There are various types of LAMP assays conventional LAMP, multiplex LAMP, in-disc LAMP, electric LAMP and reverse-transcription LAMP (Wong *et al.*, 2017). LAMP products have a distinctive ladder-like pattern, due to the variable number of inverted repeats (Tomlinson and Boonham, 2008). LAMP is not applicable for cloning purposes (Sahoo *et al.*, 2016), complicated primer design (Lau and Botella, 2017). LAMP is used to detect plant viruses, oomycetes fungi, and plant-parasitic nematodes. The process involves the use of several primers and a DNA polymerase with strand displacement activity.

Materials Required

- DNA sample containing the target sequence
- LAMP primers (inner primers, outer primers, and loop primers)
- Bst DNA polymerase or a similar strand displacement DNA polymerase
- Reaction buffer containing Mg^{2+} and other necessary components for the enzyme
- dNTPs (deoxynucleotide triphosphates)
- Incubator or heat block set to a constant temperature (typically 60-65°C)

Procedure

Design Primers: Design specific LAMP primers that target the region of interest in the DNA sequence. LAMP primers consist of six sequences (F1, F2, B1, B2, LoopF, and LoopB), and their design requires a bioinformatics tool or software to ensure specificity.

Prepare LAMP Reaction Mix: In a sterile microcentrifuge tube, prepare the LAMP reaction mix by combining the following components:

a) DNA sample (containing the target sequence) or a template control (water if no DNA is to be used).

b) Forward inner primer (FIP) and backward inner primer (BIP) at a final concentration of around 1-2 μM each.

c) Forward outer primer (F3) and backward outer primer (B3) at a final concentration of around 0.2-0.4 μM each.

d) Forward loop primer (LF) and backward loop primer (LB) at a final concentration of around 0.2-0.4 μM each.

e) dNTPs at a final concentration of around 0.2-0.8 mM each.

f) Bst DNA polymerase or similar strand displacement DNA polymerase at the recommended concentration.

g) Reaction buffer containing Mg^{2+} at the appropriate concentration for the enzyme.

Mix and prepare reaction tubes: Thoroughly mix the reaction components by pipetting or vortexing.Divide the reaction mix into separate tubes according to the number of reactions needed (e.g., samples, controls, and standards).

Incubate at Constant Temperature: Place the reaction tubes in an incubator or a heat block set to a constant temperature of around 60-65°C (depending on the specific DNA polymerase used) for typically 1 to 2 hours.

Visualize Results (Optional): After the incubation period, you can visualize the LAMP results by using various methods such as gel electrophoresis, colorimetric indicators, or turbidity-based methods. These methods can help determine the presence or absence of the target DNA.

Data Analysis: Analyse the LAMP results, interpreting the presence or absence of the amplified DNA and comparing it to positive and negative controls.

Recombinase polymerase amplification (RPA)

RPA is another isothermal DNA amplification and detection technique that requires the use of primers, nucleotides, DNA recombinase, a strand displacement polymerase enzyme and single-stranded DNA binding proteins. RPA depends on an enzymatic activity to separate the dsDNA in order to facilitate primer binding to the target sequences, hence it does not require an initial denaturation step to denature the target DNA (Piepenburg *et al.*, 2006; Labato and O' Sullivan, 2018).

Principle of RPA

RPA is based on the use of recombinase enzymes that facilitate the binding of primers to their target sequences on the DNA or RNA template. The recombinase creates a DNA strand displacement, allowing the primers to initiate the amplification process. The reaction occurs at a constant temperature (usually between 37-42°C), eliminating the need for a thermal cycler.

Steps

1. The recombinase proteins bind to the primers, a complex is formed that searches the DNA template for the complementary (target) sequence.
2. Once a match is identified, the recombinase-primer complex adheres to the complementary sequence by strand invasion (37-42°C), which results in the complementary strand displacement.
3. The complex will be stabilised by the ssDNA binding proteins in order to avoid the displaced strand from re-hybridizing with the template DNA.
4. In the presence of dNTPs, the strand displacement polymerase extends (37-42°C) the hybridised primer. The same process takes place in the reverse direction and the target DNA is exponentially amplified. Detection is done either by a lateral flow device or through gel electrophoresis.

Advantages of RPA for plant disease detection

Speed: RPA can produce results in as little as 10-30 minutes, making it a rapid diagnostic tool for plant disease detection.

Sensitivity: RPA is highly sensitive and can detect low amounts of target DNA or RNA, even in complex plant samples.

Simplicity: The isothermal nature of RPA and the minimal equipment requirements make it suitable for use in resource-limited settings and field conditions.

Specificity: RPA primers can be designed to target specific pathogen sequences, ensuring high specificity for the detection of particular plant pathogens.

It is one of the fastest methods for nucleic acid amplification, the process typically completes in 20 minutes. RPA can be paired with a reverse transcription to target RNA templates for the purpose of RNA virus identification. The limitation of RPA is primer mismatching in case of similar DNA sequences which ultimately leads to false positive results (Babul *et al.*, 2018). RT-RPA has been used to detect plant viruses such as little cherry virus (Mekuria *et al.*, 2014) and plum pox virus (Zhang *et al.*, 2014).

HDA (Helicase Dependent Amplification)

One of the simplest techniques for isothermal nucleic acid amplification, HDA mimics the in vivo process of DNA replication by isothermally unwinding DNA duplexes rather than heating them as in the case of PCR. HDA technique was developed by New England Biolabs (Vincent *et al.*, 2004). Although this isothermal technique is quite similar to a traditional PCR, it does not rely on heat denaturation for separating the double-stranded DNA and it enables primers to anneal to their complementary target sequences. For primer annealing and primer extension under isothermal conditions, HDA generates single-stranded DNA via DNA helicase. Two additional proteins, MutL and a single-stranded DNA-binding protein (SSB), are needed in addition to DNA helicase to unwind double-stranded DNA in order to stabilise single-stranded DNA and prevent the re-hybridization of the complementary strands. Two oligonucleotide primers hybridise to the 5′ and 3′ borders of the target sequence shortly after DNA has been isothermally unwound and stabilised. A DNA polymerase then extends the annealed primers by adding deoxynucleotide-triphosphates (dNTPs), resulting in double-stranded amplification products.

The dNTPs mixture should consist of Mg^{2+} which is a cofactor for both polymerase and helicase while dNTP is a cofactor of helicase. The coordination of helicase and polymerase activity is crucial. A thermostable UvrD helicase (Tte-UvrD) from *Thermoanaerobacter tengcongensis* that can be amplified at a higher temperature has been discovered, which represents an important breakthrough. HDA may be carried out at a higher temperature because to the use of Tte-UvrD, which also eliminates the requirement for SSB and MutL by reducing the re-annealing of single-stranded DNA.HDA is used to detect tobacco mosaic virus, citrus leprosis virus and bacteria, Staphylococcus and Streptococcus. HDA relies on a helicase enzyme that unwinds the double-stranded DNA, creating single-stranded templates for primer binding. A DNA polymerase then synthesizes new strands, leading to exponential amplification of the target DNA.

Steps involved in HDA for plant disease detection

1. Sample collection and preparation: Collect a plant sample suspected of harbouring the pathogen. Prepare the sample to extract the nucleic acids, which may include DNA or RNA, depending on the pathogen type.
2. Primer design: Design specific primers that recognize conserved regions of the pathogen's nucleic acid sequence. These primers will be used to initiate the amplification process.
3. HDA reaction setup: Mix the extracted nucleic acids with the designed primers, a DNA helicase enzyme, and a DNA polymerase with strand displacement activity. Additionally, include reaction buffer and necessary co-factors.
4. Isothermal amplification: The reaction is incubated at a constant temperature, typically between 60°C to 65°C, which allows the helicase enzyme to unwind the DNA and the DNA polymerase to amplify the target

sequence. The isothermal nature of the reaction eliminates the need for a thermal cycler.

5. Detection: Various methods can be employed to detect the amplified product. Some common methods include agarose gel electrophoresis, fluorescence-based assays, lateral flow devices, or real-time fluorescence monitoring.

Benefits of HDA for plant disease detection

Rapid results: HDA is an isothermal technique that can provide results within a short period, usually less than an hour.

Portability: Its isothermal nature allows it to be adapted for use in the field with portable equipment, making it suitable for on-site diagnostics.

Sensitivity and specificity: When designed properly, the primers used in HDA can provide high specificity, ensuring accurate detection of the target pathogen.

Cost-effective: The isothermal conditions and simplified setup reduce the need for expensive equipment, making it a cost-effective alternative to PCR-based methods.

CRISPR-Cas-Based Assays: The Clustered Regularly Interspaced Short Palindromic Repeats (CRISPR)-Cas system can be adapted for specific detection of plant pathogens. CRISPR-Cas-based assays offer rapid and sensitive pathogen detection. There are two main types of CRISPR-Cas-based assays used for plant pathogen detection:

i) CRISPR-Cas9-based assays

CRISPR-Cas9 is the most well-known and widely used CRISPR system. In this approach, the Cas9 protein is guided by a short RNA molecule called a single-guide RNA (sgRNA) to a specific target DNA sequence of the plant pathogen. If the target sequence is present, Cas9 will cleave the DNA, resulting in a double-stranded break. This cleavage event can then be detected through various methods, such as fluorescence-based assays, lateral flow assays, or polymerase chain reaction (PCR)-based methods.

ii) CRISPR-Cas12-based assays (also known as CRISPR-Cas12a or CRISPR-Cpf1)

CRISPR-Cas12 is another CRISPR system that has gained attention for its application in diagnostics. Similar to CRISPR-Cas9, Cas12 is guided by a CRISPR RNA (crRNA) to recognize the target sequence. Upon target binding, Cas12 becomes activated and exhibits collateral cleavage activity, indiscriminately cleaving nearby non-target ssDNA molecules. This collateral cleavage can be exploited to detect the presence of the target sequence. Fluorescence-based reporters or lateral flow assays are commonly used to visualize the cleavage event. Both CRISPR-Cas9 and CRISPR-Cas12-based assays can be combined with different signal amplification strategies to increase sensitivity. For example, the use of isothermal amplification techniques like recombinase polymerase amplification (RPA) or loop-mediated isothermal amplification (LAMP) can enhance the detection limit of the assay.

The workflow of a typical CRISPR-Cas-based assay for plant pathogen detection involves the following steps:

- **Sample collection:** Obtain a sample from the plant suspected of being infected with the pathogen.
- **Nucleic acid extraction:** Isolate and purify the DNA or RNA from the sample. This step is crucial for obtaining high-quality genetic material for subsequent analyses.
- **Amplification (optional):** If the target pathogen's genetic material is present in low abundance, a nucleic acid amplification step might be necessary to increase the detection sensitivity.
- **CRISPR-Cas detection:** Combine the extracted nucleic acids with the CRISPR-Cas components (Cas protein and guide RNA) that specifically target the pathogen's DNA or RNA.
- **Signal readout:** Analyse the reaction output using fluorescence-based methods, lateral flow assays, or other detection methods. The presence of a specific signal indicates the presence of the target pathogen.

Next-Generation Sequencing (NGS): NGS technologies enable the sequencing of millions of DNA or RNA fragments in parallel. They can be used for metagenomic analyses of plant samples to identify multiple pathogens simultaneously. Steps for NGS are as follows:

- **Sample Collection:** The first step involves obtaining a biological sample from the patient or environment suspected of harbouring the pathogen. This could be blood, tissue, swabs, or other relevant specimens.
- **Nucleic Acid Extraction:** NGS requires high-quality DNA or RNA, so the next step is to isolate and extract the genetic material (DNA or RNA) from the collected samples.
- **Library Preparation:** In this step, the isolated genetic material is converted into a sequencing library. The library contains small fragments of the DNA or RNA, which are tagged with unique molecular identifiers, making it easier to identify and quantify the different sequences during data analysis.

- **Sequencing:** The prepared library is then loaded onto the NGS platform for sequencing. This process generates vast amounts of short DNA sequences (reads) from the genetic material in the library.
- **Data Analysis:** The raw sequencing data is processed and analysed to identify the presence of pathogens. Bioinformatics tools are used to align the short reads to reference genomes or perform de novo assembly to reconstruct the pathogen's genome. The data is then compared to databases of known pathogens to determine which organisms are present in the sample.
- **Pathogen Identification:** The final step involves the identification of pathogens based on the genomic data obtained. This could include bacteria, viruses, fungi, or other infectious agents. The information obtained from NGS can also reveal important characteristics of the pathogens, such as drug resistance mutations and virulence factors.

Advantages of NGS for pathogen detection

- **Comprehensive:** NGS can simultaneously detect multiple pathogens in a single sample, making it a valuable tool for detecting co-infections and mixed infections.
- **Sensitivity:** NGS can detect low-abundance pathogens, which might be challenging with traditional diagnostic methods.
- **Speed:** NGS enables rapid detection and identification of pathogens, reducing the time required for diagnosis.
- **Surveillance and Outbreak Investigation:** NGS allows for real-time monitoring of pathogen evolution and spread, aiding in the investigation of outbreaks and transmission dynamics.
- **Personalized Medicine:** NGS can provide insights into an individual's pathogen profile, enabling tailored treatment strategies based on the detected pathogens' genetic makeup.

Remote Sensing and Drones: Remote sensing using satellite imagery, drones, or unmanned aerial vehicles (UAVs) allows for large-scale monitoring of crops. Hyperspectral sensors on drones can detect subtle changes in plant reflectance, indicating stress or disease presence. The data collected can be analysed using machine learning algorithms to identify disease patterns.

Key techniques used in remote sensing for plant disease detection are:

a) **Multispectral and Hyperspectral Imaging:** Multispectral and hyperspectral sensors capture data in multiple narrow or contiguous bands across the electromagnetic spectrum. This allows for the detection of subtle variations in plant health, including early signs of disease. Hyperspectral imaging provides even more detailed spectral information, enhancing disease identification accuracy.

b) **Thermal Imaging:** Thermal cameras detect variations in plant temperature, which can indicate stress caused by diseases. Infected plants often exhibit different thermal profiles compared to healthy ones, enabling the identification of diseased areas.

c) **LiDAR (Light Detection and Ranging):** LiDAR sensors use laser pulses to measure the distance between the sensor and the target surface, creating high-resolution 3D models of the vegetation. This technology can be used to monitor changes in plant structure and health, which may be indicative of disease presence.

d) **Unmanned Aerial Vehicles (UAVs or Drones):** Drones equipped with remote sensing instruments can fly over fields and capture high-resolution images and data, allowing for rapid and frequent monitoring of large areas. They provide valuable information on disease spread patterns and can target specific locations for disease control measures.

e) **Satellite Remote Sensing:** Satellites equipped with various sensors can monitor vast agricultural areas from space. Satellite imagery allows for frequent and consistent monitoring, which is especially useful for large-scale disease outbreak detection and tracking.

f) **Machine Learning and Image Analysis:** The collected remote sensing data is processed using advanced image analysis techniques and machine learning algorithms. These algorithms can learn to recognize patterns associated with diseased plants and classify them accurately, leading to automated disease detection and mapping.

g) **Spectral Indices:** Spectral indices, such as the Normalized Difference Vegetation Index (NDVI) and the Normalized Difference Water Index (NDWI), are derived from remote sensing data and can indicate plant health and stress, which may be linked to disease presence.

Benefits of Remote Sensing for Plant Disease Detection

- **Early Detection:** Remote sensing allows for the early detection of diseases before they become visually apparent. This early warning can help implement timely interventions and minimize crop losses.
- **Precision Management:** Remote sensing data can be used to create detailed disease maps, enabling farmers to target specific areas for treatment rather than applying treatments to entire fields.
- **Cost-Effective:** By identifying and targeting diseased areas, farmers can reduce the unnecessary use of pesticides and other inputs, leading to cost savings and reduced environmental impact.

- **Large-Scale Monitoring:** Remote sensing covers vast areas, making it efficient for monitoring large agricultural regions, including remote and inaccessible locations.
- **Real-Time Monitoring:** With advances in technology, remote sensing systems can provide real-time or near-real-time data, enabling prompt responses to disease outbreaks.
- **Machine Learning and Artificial Intelligence:** Machine learning algorithms, particularly deep learning techniques, have proven effective in analysing complex data sets, such as images and spectral data, to identify patterns associated with diseased plants. Models like convolutional neural networks (CNNs) can be trained on vast datasets to achieve high accuracy in disease detection.

Nanotechnology

Nanotechnology has shown great promise in various fields, including agriculture. When it comes to plant disease detection, nanotechnology offers innovative solutions that can improve accuracy, sensitivity, and real-time monitoring. Here are some ways nanotechnology can be used for plant disease detection:

Nanosensors: Nanosensors are tiny devices capable of detecting specific biological or chemical substances. In plant disease detection, nanosensors can be designed to identify disease-causing pathogens or the biochemical markers produced by infected plants. They can be integrated into plant tissues, soil, or even carried by drones to scan large agricultural areas. Nanosensors enable rapid and precise detection, allowing farmers to respond promptly to disease outbreaks.

Nanomaterial-based assays: Nanomaterials, such as quantum dots, gold nanoparticles, or carbon nanotubes, can be used in diagnostic assays for plant diseases. These nanomaterials can be functionalized with specific antibodies or nucleic acid probes, allowing them to bind selectively to the target pathogens or biomarkers. When the target molecules are present, the nanomaterials emit characteristic signals that can be easily detected, providing a simple and sensitive detection method.

Conclusion

More advanced methods for detecting plant diseases have been developed recently as a result of technological improvements. A quick, sophisticated, and on-site testing technique called a point-of-care (POC) diagnostic assay for plant disease detection is used to determine whether pathogens are present in plants. This kind of diagnostic tool is especially helpful Plant viruses, oomycetes fungus, and plant-parasitic nematodes can all be identified using loop-mediated isothermal amplification (LAMP). To boost sensitivity, CRISPR-Cas9 and CRISPR-Cas12-based assays can be paired with various signal amplification techniques. In the presence of the target sequence, Cas9 will cleave the DNA, causing a double-stranded break. Then, using a variety of techniques, such as lateral flow tests, fluorescence-based assays, or PCR-based techniques, this cleavage event can be observed. Remote sensing techniques can be used to monitor and identify plant diseases from a distance, without direct physical contact with the plants. So that farmers can gain valuable insights into the health of their crops, identify diseases at early stages, and take appropriate actions to mitigate the spread and impact of plant diseases. Nanosensors can be used to detect infections that cause disease or the biochemical indicators that sick plants produce.

References

Babu, B., Ochoa-Corona, F.M. and Paret, M.L. (2018). Recombinase poly-merase amplification applied to plant virus detection and potential implications. Analytical Biochemistry, 546: 72-77

Chen, H.W., Weissenberger, G., Atkins, E., Chao, C.C., Suputtamongkol, Y. and Ching, W.M. (2015). Highly sensitive loop-mediated isothermal amplification for the detection of Leptospira. International Journal of Bacteriology, pp: 201

Chow, W.H.A., McCloskey, C., Tong, Y., Hu, L., You, Q., Kelly, C.P., Kong, H., Tang, Y.W. and Tang, W. (2008). Application of isothermal helicase-dependent amplification with a disposable detection device in a simple sensitive stool test for toxigenic Clostridium difficile. The Journal of Molecular Diagnostics, 10(5): 452-458

Craw, P. and Balachandran, W. (2012). Isothermal nucleic acid amplification technologies for point-of-care diagnostics: a critical review. Lab Chip, 12(14): 2469–2486

Feng, W., Hieno, A., Kusunoki, M., Suga, H. and Kageyama, K. (2019). LAMP detection of four plant-pathogenic oomycetes and its application in lettuce fields. Plant Disease, 103(2): 298-307

Gill, P. and Ghaemi, A. (2008). Nucleic acid isothermal amplification technologies – A review. Nucleosides Nucleotides Nucleic Acids, 27(3): 224–243

Goto, M., Honda, E., Ogura, A., Nomoto, A. and Hanaki, K.I. (2009). Colorimetric detection of loop-mediated isothermal amplification reaction by using hydroxy naphthol blue. Biotechniques, 46(3): 167-172

Iwamoto, T., Sonobe, T. and Hayashi, K. (2003). Loop-mediated isothermal amplification for direct detection of Mycobacterium tuberculosis complex, M. avium, and M. intracellulare in sputum samples. Journal of Clinical Microbiology, 41(6): 2616-2622

Lau, H.Y. and Botella, J.R. (2017). Advanced DNA-based point-of-care diagnostic methods for plant diseases detection. Frontiers in Plant Science, 8: 2016

Lobato, I.M. and O'Sullivan, C.K. (2018). Recombinase polymerase amplification: Basics, applications and recent advances. Trac Trends in Analytical Chemistry, 98:19-35

Lutz, S., Weber, P., Focke, M., Faltin, B., Hoffmann, J., Müller, C., Mark, D., Roth, G., Munday, P., Armes, N. and Piepenburg, O. (2010). Microfluidic lab-on-a-foil for nucleic acid analysis based on isothermal recombinase polymerase amplification (RPA). Lab on a Chip, 10(7): 887-893

Mahalanabis, M., Do, J., Almuayad, H., Zhang, J. Y. and Klapperich, C. M. (2010). An integrated disposable device for DNA extraction and helicase dependent amplification. Biomedical Microdevices, 12: 353–359

Mekuria, T.A., Zhang, S. and Eastwell, K.C. (2014). Rapid and sensitive detection of Little cherry virus 2 using isothermal reverse transcription-recombinase polymerase amplification. Journal of Virological Methods, 205: 24-30

Mori, Y., Kitao, M., Tomita, N. and Notomi, T. (2004). Real-time turbidimetry of LAMP reaction for quantifying template DNA. Journal of Biochemical and Biophysical Methods, 59(2):145-157

Mori, Y., Nagamine, K., Tomita, N. and Notomi, T. (2001). Detection of loop-mediated isothermal amplification reaction by turbidity derived from magnesium pyrophosphate formation. Biochemical and Biophysical Research Communications, 289(1): 150-154

Notomi, T., Okayama, H., Masubuchi, H., Yonekawa, T., Watanabe, K., Amino, N. and Hase, T. (2000). Loop-mediated isothermal amplification of DNA. Nucleic Acids Research, 28(12): e63-e63

Piepenburg, O., Williams, C.H., Stemple, D.L. and Armes, N.A. (2006). DNA detection using recombination proteins. PLoS Biology, 4(7): e204

Rohrman, B. A. and Richards-Kortum, R. R. (2012). A paper and plastic device for performing recombinase polymerase amplification of HIV DNA. Lab Chip, 12(17): 3082–3088

Sahoo, P.R., Sethy, K., Mohapatra, S. and Panda, D. (2016). Loop mediated isothermal amplification: An innovative gene amplification technique for animal diseases. Veterinary World, 9(5): 465

Tomlinson, J. and Boonham, N. (2008). Potential of LAMP for detection of plant pathogens. CABI Reviews, 1-7

Vincent, M., Xu, Y. and Kong, H. (2004). Helicase-dependent isothermal DNA amplification. EMBO Reports, 5(8): 795-800.

Wong, Y.P., Othman, S., Lau, Y.L., Radu, S. and Chee, H.Y. (2018). Loop-mediated isothermal amplification (LAMP): a versatile technique for detection of micro-organisms. Journal of Applied Microbiology, 124(3): 626-643

Yager, P., Edwards, T., Fu, E., Helton, K., Nelson, K., Tam, M.R. and Weigl, B.H. (2006). Microfluidic diagnostic technologies for global public health. Nature, 442(7101): 412-418

Zhang, S., Ravelonandro, M., Russell, P., McOwen, N., Briard, P., Bohannon, S. and Vrient, A. (2014). Rapid diagnostic detection of plum pox virus in Prunus plants by isothermal AmplifyRP® using reverse transcription-recombinase polymerase amplification. Journal of Virological Methods, 207: 114-120

2

Modern Technologies of Insect Trapping

Ashutosh Singh Aman, Arun Kumar, Pramod Kumar Mishra, Rishabh Mishra, Paritosh Tripathi and Pankaj Kumar Rajpoot

Department of Entomology, Chandra Shekhar Azad University of Agriculture & Technology, Kanpur-208002 (U.P.), India

Introduction

There are multiple methods or trapping technologies available for pest management in agriculture. Some of the method discussed in literature are described here. Insect pests in an agricultural field can cause damage to the crops by infecting the crops with diseases and thus may affect adverse impact on its quality. Nowadays, farmers set up insect traps like sticky papers that comes in different colours such as yellow, green and blue to monitor the number of pests. Different insects are specifically/ particularly attracted to the colours of the sticky paper commonly such as yellow or blue (Espinoza *et al.*, 2016). Using sticky papers farmers can estimate the pest density trying to damage the crops. But for very small insects, the counting is very tedious and time-consuming and may be inaccurate at times.

Pests and insects are drawn to plants because they emit volatile biological amalgams. Plants have their own self-defence mechanisms, and their reaction to insect attacks differs based on a variety of damaging factors. As a result, pest management has become an integral aspect of crop protection. Induced resistance is a prominent strategy in insect management strategies. This study is extremely useful in genetically developing plants to produce adequate and necessary self-protective chemicals (War *et al.*, 2012).

Because ultraviolet (UV) rays are visible to insects, they are easily drawn to them. To catch daylight insects, yellow pan traps with yellow sticky plates are used. Lamps with yellow illumination are also used at night to suppress pests. As a result, the extent of insect pest harm done to flowers, vegetables, and fruits can be limited. Hovering bug populations can be reduced by covering agricultural farms or fields with reflective materials (Shimoda and Honda, 2013).

In a pest control system that uses a wireless sensor grid, a scattered imaging device is incorporated. A host station, a master node, and a few client nodes are included in this design. The caught portion of the trapped insects is collected as a picture and relayed to the host station. The station includes a reference insect density. When the original insect density surpasses the reference density, an alarm is generated and sent to client nodes via the master node. A ZigBee transceiver is used to convey pest information among nodes, which is then distributed to farmers through message or phone call (Priya *et al.*, 2013). The traditional sensor-node trapping method is modified with current web server capable sensors for dynamic response. Transmission at high speeds is guaranteed. This approach is also useful for remote field access. An agent program is used to evaluate the same, and the investigative results show that the given strategy is superior (Fukatsu and Hirafuji, 2005).

The Indian Agricultural Research Institute, New Delhi, conducted field research at regular intervals to find and analyse the efficacy of various light sources in light traps used to check many pest populations. The experiment made use of mercury lamps, black lights, and ultra violet lights. Mercury light trap had the greatest capability of the three light traps, followed by Black light trap and UV light trap (Ramamurthy *et al.*, 2010). In Australia, a wireless sensor design is being utilized to protect the fruit and vegetable production from fruit fly. Fruit fly detection and recognition algorithms use machine vision approaches. The average precision of this system has been determined experimentally to be 80% (Liu *et al.*, 2009).

To test the picture transfer rate, an image sensor network is created and compared to ZigBee. The JPEG images are verified, and the results show that the encoded images are more error resistant since they have numerous layers (Pekhteryev *et al.*, 2005). The efficacy of the trap is determined on the time of insect attack and the location of the bug. Predicting them in a manual trap is time-consuming. To improve the efficiency of the existing approaches, a sensor is added and insect recording is enabled. An insect detecting sensor comprised of phototransistors is developed. These photo gadgets are illuminated by infrared LEDs. The insects are discovered and recorded using wing strokes. The data is transmitted from the field to the smartphone (Potamitis *et al.*, 2014; Kim *et al.*, 2008).

Pests and diseases cause great economic loss to farmers through reduced yields and increased costs of pesticides and other control measures. Thereby use of different technologies of insect trapping yields both economic and environmental benefits. Environmental issues about pesticide usage, overexploitation of natural resources, changes in consumer habits, and advancements in technology are growing, bringing in a new agricultural revolution. This revolution involves the use of digital tools to boost productivity while also optimizing the management of natural resources and agricultural inputs (Fraser *et al.*, 2019).

Different methodologies of insect trapping for controlling of crop attacking insects

In the past, insecticides were employed to control the insects. Though they were effective in repelling insects, the quality of the harvest was poor. Furthermore, the fertility of the ground was harmed. Researchers are currently striving to restore soil quality through ecological farming. As a result, controlling approaches shouldn't put at risk field quality or yield.

Growing a trap crop is one of the most basic pest management approaches. It is an attractive host plant that attracts attention and keeps insects away from the main crop throughout the considerable period of main crop. Pests and flies are more interested in the reproductive periods of host plants than in the vegetative phases. This principle is used in trap cropping to control pests. Some of the benefits of trap cropping include increased yield quality, the attraction of beneficial species, increased biodiversity, and reduced insecticide use. This strategy, however, does not provide a complete answer to all types of existing insect infestations. Rather, successful pest management necessitates knowledge and understanding of insect behaviour.

Insect traps are the next step in pest management. They are used to monitor and reduce insect populations by employing baits such as food, visual lures, chemical attractants, and pheromones. The trap mechanism or bait varies depending on the crop or insects. The food traps are not meant to harm any animals. Food traps may attract a variety of insects. Proteins are attracted to flies and wasps. Lactic acid, flowery, and fruity scents attract mosquitos and other insects. Visual lures come in a variety of vibrant colours and shapes. The strength and colour of light vary depending on the insect being trapped. Chemical attractants may just attract one sex. Certain insects are drawn to light traps with or without ultra violet light. The quantity of insects attracted by these traps may vary depending on environmental factors such as temperature and humidity. Flying and ground insects can be attracted by light traps.To catch diverse insects, a mix of light traps and other traps is utilized. Indoor pest monitoring employs sticky traps.

Because most traps are UV sensitive, only ultra violet traps are employed in the current system of insect traps. In the existing systems, there are no modern picture capture techniques. Furthermore, no timers or counters are used in the current systems. The recording of visual information and statistical data is critical in pest management. As a result, in this chapter, a technique for incorporating modern recording systems is presented.

Types of insects capturing trap

The overview of the following fundamental trap types, as well as modifications of those basic or common trap, types for various applications. Traps can be setup with or without attractant cues, and they can use a mix of visual (light, shape, colour), chemical (oviposition, pheromone, parapheromone, food/host,), and audio signals to make traps more specific and effective. There are automated or self-monitoring devices that will send information to trap catch to an outside station. Traps of insects are a critical component of Insect Pest Management (IPM) systems. Though several systems of trapping or trapping system are addressed, the goal is to provide a broad framework of trap types with some illustrative examples. It is not aim or intended to be an exhaustive list of all traps discovered. However, different traps are utilized for insects that move through the air (wind-borne or flying insects), soil-dwelling or walking insects, aquatic and subterranean insects. Several of these are discussed more below.

Traps for Windblown insects and other flying insects

Interception trap

This interception traps are often employed for faunal/insect surveys in ecological studies, but they can also be utilized for management of pest. It is a hanging net with a top penetration that leads to a collecting tube in its most basic form. An example of an interception trap is the Malaise trap. Insect migration has been studied using fixed interception traps with various collecting tubes for north-bound versus south-bound insects. Large funnel-shaped nets have been put on moving trucks to collect flying insects like biting midges (Ceratopogonidae) or coupled to a suction mechanism to collect Russian wheat aphids, *Diuraphisnoxia* (Mordvilko).

Sticky traps

Panels, cylinders, or spheres coated with sticky substance are undoubtedly the most widely utilized traps in agricultural studies for faunal surveys. In their most basic form, Trap may be like panels coated or applied with substances that will trap insects that are blown onto the panel or fly into it. Panels can also be combined with a colour, a shape, and a chemical attractant. A thin coating of motor oil will keep very small insects in place, but most often larger insects get escaped. Sticky material such as Tangle-trap is applied to the surface to capture both small and large insects.

3-dimensional triangular trap, Diamond traps and Wing traps

This is the other type of trap of low-cost that are employed in conjunction with an attractant. These are essentially a sticky trap with the adhesive surface covered on the interior of the trap, and they are both disposable and utilized with a sticky or adhesive component that is replaced after servicing. Trap known by name Delta trap are compact and light triangle (tent-shaped) traps that are easy to hang/suspend and transport. The Jackson trap is a type of delta trap that is used to catch a variety of *tephritid* fruit flies. It includes a removable base that is coated with sticky or adhesive substance and may be replaced as necessary during sampling. In public places such as stores and supermarkets, diamond-shaped traps are used to monitor interior pests. For pest Lepidoptera, wing traps with a ceiling and a floor are utilized.

Bucket traps

A bucket trap is another simple and low-cost trap. This trap can be used without a lid or as a closed container with provided holes for insects to enter on the top or sides. Small drainage holes near the bottom side of the trap can be used to allow trapped water to drain out. Bucket traps, such as the Nadel trap and the Mission trap, are translucent or semi-transparent, and closed containers with hole of entrance around the edge. The corn rootworm trap has a huge dome-shaped top with a gap between the top and the bottomas well as the container. The targeted insect determines the size and placement of the container, as well as the diameter and number of entry holes. These are typically used in conjunction with a chemical lure of some kind. These can be used without a lethal chemical to retrieve live insects, however caught insects may escape via the entrance openings. Bucket-type traps can be utilized with a variety of retention devices. Inside, a pesticide such as dimethyl 2, 2-dichlorovinyl phosphate (DDVP, mothballs) that can be applied.

Cone trap

This cone traps are similar to bucket traps with funnels; however, the funnel is much larger in comparison to the bucket or collecting tube. These are utilized with the funnel's big aperture facing the ground and the funnel's top heading up to a container. *Heliothis* traps and butterfly bait traps are two examples of these traps.

Traps for soil- dwelling insects and other walking arthropods Pitfall trap

This pitfall traps are effective for catching arthropods and other insects as they travel across a surface. This surface is often a ground or soil surface for capturing spiders, beetles and other ground-dwelling species, though they may also catch/seize flying insects as they move around the soil surface. Pitfall traps put on the top of stored grain have also been utilized to collect insect samples. Pitfall traps are usually made up of a container buried inside the ground substrate into which arthropods and other insects falls and gets trapped.

Grain Probe trap

Grain probe traps are a type of pitfall trap that has been modified for usage in stored grain. Grain probe traps are made up of an elongated cylinder along with drilled holes into the sides and above a funnel and receptacle of insects.

Shelter trap

Shelter traps are used for insects that harbourage in dark and are effective in situations where the trap must be hidden. These are typically used to catch insects as they walk or move across a surface, and they include small apertures that allow insects to enter the trap. The well-known instances of such traps are roach motels. They usually contain an adhesive material inside, to trap insects that enters or an oil or other chemical to keep attracted insects at bay. Shelter traps are used to catch stored goods insects and come in a variety of designs such as dome traps, stealth traps, and corner traps. They provide nesting site for Bees.

Emergence trap

Emergence traps are cone traps designed to catch adults with underground larval and pupal phases. These traps are made up of aluminium screening bent into the shape of a funnel, with the top aperture leading or forming to a collecting tube. To seal the trap to the earth, the trap is positioned the earth aligns with the surface, and soil is being pushed up around the edges. Adults emerge from belowground stages and enter the trap, where they are captured in the collecting tube.

Solar bait stations

Solar bait stations, on the other hand, can capture the underground stages of wireworms (Elateridae) and fake wireworms (Tenebrionidae). These are used to estimate wireworm numbers and determine seed treatment/non-treatment decisions. These insect larvae can be aggregated by establishing micro-environments with favourable temperature, moisture and feeding.

Traps for aquatic insects

Interception trap

Interception traps collect aquatic insects as they move through the water and are comparable to wind-borne insect interception traps. These traps are frequently becoming narrow nets with round or rectangular apertures that can be placed in place to sample insects in running water or manually moved through the water. The size of the mesh determines the size of the insects kept; however very fine mesh may limit water circulation through the trap while retaining abundant trash.

Emergence trap

When aquatic insects with larval stages emerge as young one or adults, floating emergence traps are employed to capture them. Construction and operation are similar to that of soil-dwelling insect emergence traps. Communication and assessment of their surroundings are important. In insect traps, Semiochemicals with a behavioural influence on insect orientation or direction, that is, chemicals that cause an insect to move towards the source are utilized.

Traps for Capturing Insects, Fig. 1: Insect traps such as: first row left, Malaise trap, an interception trap; first row right, Steiner trap, a lure-based trap; second row left, PC floor trap, a lure-based trap; second row right, pitfall trap; third row left, solar bait station; third row right, suspended black ball coated with adhesive substance; fourth row left, red sphere on yellow panel; fourth row right, boll weevil trap.

Traps for Capturing Insects, Fig. 2: Some additional insect traps: first row left, grain probe traps top right, blacklight trap; second row left, wing trap, a lure-based trap with a sticky interior;second row right, wasp trap, a food lure-based trap; third row left, unitrap, a popular pheromone-based trap;third row right, a bucket trap;fourth row left, a cylindrical yellow panel; fourth second from left, new Jersey light trap;fourth row third from left, mosquito emergence trap; fourth row right, McPhail trap, a food lure-based trap.

Automation trap or electronic trapping

Advancement in information and computer technology and remote sensing are expanding the field of precision agriculture, which refers to the use of computers to assist in management choices. The capacity to quickly transfer data obtained from traps into computer databases or spreadsheets is an essential component of precision agriculture and will aid in making decision timely for pest management.

Analysing the insect's wingbeats coming into a trap or flying above or overhead is one potential way for automating the insect identification procedure. In feasibility studies with aphids and mosquitoes, furrier transformations, closest neighbour classification, and artificial neural network classification schemes were investigated, and classification of species accuracy were reported up to 70-90%.

An insect trap is designed with three layers of varying thickness to collect insects of various sizes, and the layers are covered in the proper bait to draw the insects. Using black lights (Ultra Violet) and LED lights, an autonomous monitoring system is ensured using a low-cost image sensor to record the photos of pests caught in the system. The battery is charged using a 10W solar panel to power the UV and LED lights using a battery. The energy produced by solar panels is kept in a battery that powers the system at night. Utilizing a solar charge controller ensures efficient charging prolongs the life of the battery and increases its capacity. The photographs that were captured are transferred to a remote-control station. The knowledge collected through this method is used to estimate the insect populations in farms. It is used in rice, brinjal and other field crop including in an open field for several insect pests such as Dragon fly, Cricket, Grasshopper, Flies, Bugs etc.

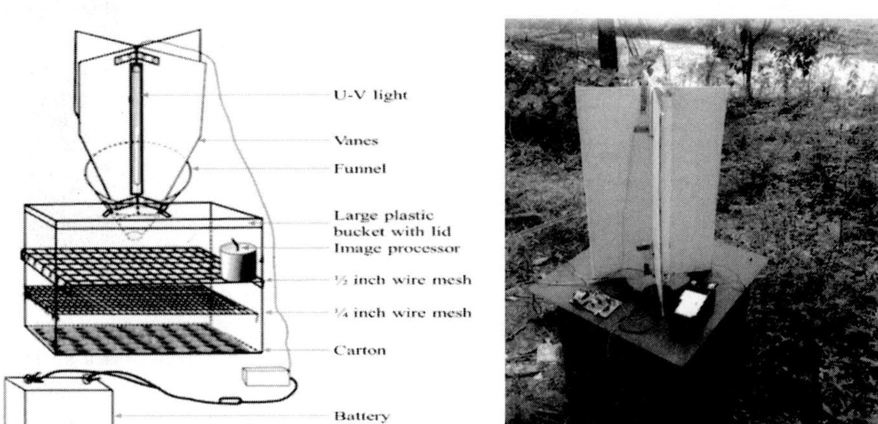

Fig. 3: Electronic trap design (in each layer of the meshes insect baits are used)

Fig. 4: Insects caught in different layers of mesh of electric trap photograph through webcam (Courtesy: MSSRF, Chennai)

Conclusion

Insect trapping systems are chief components of Integrated Pest Management (IPM) programs. Decisions can be made by trapping data about whether to initiate or discontinue control measures, as well as to evaluate the efficacy of previously adopted control procedures. The use of detection trapping alerts employees to the appearance of a new insect pest in an earlier or previously pest-free area, allowing control measures to be performed in a scheduled time. Early diagnosis and targeting of tiny infestations or incidence will aid pest management efforts.

The overview of the following fundamental trap types, as well as modifications of those basic or common trap, types for various applications. Traps can be setup with or without attractant cues, and they can use a mix of visual (light, shape, colour), chemical (oviposition, pheromone, parapheromone,food/ host,), and audio signals to make traps more specific and effective. There are automated or self-monitoring devices that will send information to trap catch to an outside station. Traps of insects are a critical component of Insect Pest Management (IPM) systems. Though several systems of trapping or trapping system are addressed, the goal is to provide a broad framework of trap types with some illustrative examples. It is not aim or intended to be an exhaustive list of all traps discovered.

References

Epsky, N.D., Morrill, W.L. and Mankin, R.W. (2008). Traps for capturing insects. Encyclopedia of Entomology, 3: 2318-2329

Espinoza, K., Valera, D.L., Torres, J.A., López, A. and Molina-Aiz, F.D. (2016). Combination of image processing and artificial neural networks as a novel approach for the identification of Bemisia tabaci and Frankliniella Occidentalis on sticky traps in greenhouse agriculture. Comput. Electron. Agric. 127: 495–505

Fraser, A. (2019). Land grab/data grab Precision agriculture and its new horizons. J. Peasant Stud. 46: 893–912

Fukatsu, T. and Hirafuji, M. (2005). Field Monitoring Using Sensor-Nodes with a Web Server. J. Robotics Mechatronics, 17(2): 164-172

Hendrick, H., Liao, G. S., Lu, K. Y., Lin, C. Y. and Jong, G. J. (2016). The Trapping Device Implementation of Wireless Sensor Network. In Cloud Computing: 6th International Conference, CloudComp 2015, Daejeon, South Korea, October 28-29, 2015, Revised Selected Papers 6 (pp. 311-316). Springer International Publishing

Kim, Y., Evans, R.G. and Iversen, W.M. (2008). Remote sensing and control of an irrigation system using a distributed wireless sensor network. IEEE Transactions on Instrumentation and Measurement, 57(7): 1379-1387

Liu, Y., Zhang, J., Richards, M., Pham, B., Roe, P. and Clarke, A. (2009, September). Towards continuous surveillance of fruit flies using sensor networks and machine vision. In 2009 5th International Conference on Wireless Communications, Networking and Mobile Computing (pp. 1-5). IEEE

Pekhteryev, G., Sahinoglu, Z., Orlik, P. and Bhatti, G. (2005, May). Image transmission over IEEE 802.15. 4 and ZigBee networks. In 2005 IEEE International Symposium on Circuits and Systems (pp. 3539-3542). IEEE

Potamitis, I., Rigakis, I. and Fysarakis, K. (2014). The electronic McPhail trap. Sensors, 14(12): 22285-22299

Priya, C. T., Praveen, K. and Srividya, A. (2013). Monitoring of pest insect traps using image sensors &dspic. Int. J. Eng. Trends Tech, 4(9): 4088-4093

Ramamurthy, V. V., Akhtar, M. S., Patankar, N. V., Menon, P., Kumar, R., Singh, S. K. and Mittal, V. (2010). Efficiency of different light sources in light traps in monitoring insect diversity. Munis Entomology & Zoology, 5(1): 109-114.

Shimoda, M. and Honda, K.I. (2013). Insect reactions to light and its applications to pest management. Applied Entomology and Zoology, 48: 413-421

Thangalakshmi, S. and Ramanujan, R. (2015). Electronic trapping and monitoring of insect pests troubling agricultural fields. International Journal, 206

War, A.R., Paulraj, M.G., Ahmad, T., Buhroo, A.A., Hussain, B., Ignacimuthu, S. and Sharma, H.C. (2012). Mechanisms of plant defense against insect herbivores. Plant Signaling & Behavior, 7(10): 1306-1320

3

Biotechnological Approaches in Pest Control

Sadhana V.[1], Suriya S.[2], Santhoshini E.[3] and Gayathri K.[4]

[1]*Department of Entomology, Tamil Nadu Agricultural University, Coimbatore-641003 Tamil Nadu*

[2] *Division of Entomology, Sher-e- Kashmir University of Agricultural Sciences and Technology, Srinagar, Jammu and Kashmir-190025*

[3]*Department of Plant Pathology, University of Agricultural Sciences, Dharwad Karnataka-580005*

[4]*Division of Entomology, Sher-e-Kashmir University of Agricultural Sciences and Technology, Srinagar, Jammu and Kashmir-190025*

Introduction

Worldwide, 40% of a crop is lost to insect damage, particularly in developing countries. Some of the major insect pests belong to the Lepidoptera and Coleoptera, and their larval or grub stages cause major damage to economically important crops including tobacco, cotton, corn, sorghum, tomato, sunflower, pulses, wheat, and lucerne. Recently, some insecticides have been primarily used for agricultural pests. Over the years, most pesticides have led to insecticide resistance in major lepidopterans, a reduction in natural enemy populations, and harm to mammals and the environment. These problems have led researchers to develop new approaches, starting with transgenic *Bacillus thuringiensis* (Bt) and newer biotechnological approaches for insect pest management, including the use of transgenic plants, Gene editing (RNAi), Gene drives, and the most recent one, CRISPR-Cas9 technology. Since the increase in the commercialization of genetically modified crops, most farmers have adopted this technology in recent years and also planted biotech crops. This has led to a reduction in the use of pesticides and an additional financial gain for farmers over the last 20 years.

Two main categories can be used to classify the application of biotechnology in pest management: (i) Traditional biotechnology and (ii) Modern bio-technology

i) Traditional biotechnological approaches used in pest management

Traditional biotechnology products for pest management include tissue culture methods, bioagents, and biofertilizers.

1. Tissue Culture Techniques

The tissue culture (TC) approach offers the rapid multiplication of disease-free plants under controlled and aseptic conditions in a relatively short period. It provides a wide range of opportunities for the development, conservation, and utilization of genetic variability for the improvement of plants. These in vitro methods found extensive use in the synthesis of secondary metabolites (Alemu, 2020). Additionally, TC is used for in vitro disease-resistant plant screening.

2. Biocontrol-Agents used in pest management

The term "bio-agents" refers to a broad category of biological agents, such as microbial agents, natural enemies, parasitoids and parasites, and botanicals. These components can be used alone or combined to create effective and holistic Integrated Pest Management (IPM) strategies to sustain agricultural production (Barzman *et al.*, 2015; Stenberg, 2017). A variety of biocontrol agents and biopesticides, which are frequently used interchangeably, have been used as traditional biotechnology products for the management of pests.

Table 1. Traditional biotechnology products used for pest management

Bio nano-agents	Bio nano-insecticidal agents (Duhan *et al.*, 2017)
Microbial agents	Entomopathogenic fungi (Sultana *et al.*, 2017); Entomopathogenic nematodes (Gozel and Gozel, 2016); Entomopathogenic bacteria and Insecticidal bacteria (Saiyad, 2017)
Sterile Insect Technique (SIT)	Control of fruit flies (Haymer, 2016)
Botanicals or plant-based biocontrol agents	Semiochemicals (Saha and Chandran, 2017), Pheromones (Kirk, 2017), Attractants (Drmic *et al.*, 2017), and Push-pull technology for insect control (Pickett *et al.*, 2014)
IPM-involving biocontrol agents or biopesticides	Integrated Insect control (Gangwar, 2017)

ii) Modern biotechnology

Molecular Breeding for Pest Management

Host plant resistance (HPR) is one of the most effective preventative measures that can also be made more reliable and fast using DNA marker-based breeding techniques. DNA markers can be regularly used in various aspects of plant genome analysis, including genome fingerprinting, gene localization, genome mapping, population genetics, taxonomy, plant breeding, and diagnostics (Chauhan and Kumar, 2015).

1. Marker-Assisted Breeding

Using DNA markers that have a close relationship to the target loci as a substitute for or in addition to phenotype screening is known as marker-assisted selection (MAS). Markers are utilized in breeding, also known as marker-assisted breeding (MAB) or molecular plant breeding, in addition to the selection of the best traits (Singh,2018). Through molecular breeding techniques, certain improved cultivars or varieties developed resistance to pests and diseases (Khera *et al.*, 2013).

2. Variants of Plant Resistance to Pests

Plants' pest resistance can be developed more quickly than through conventional breeding using the molecular techniques discussed above. There are various types of resistance, including multiple, quantitative, and durable resistances. Gene stacking and gene pyramiding were used to create durable pest-resistant varieties. Pyramiding is the stacking of multiple genes that results in the simultaneous expression of several genes in a variety to produce durable resistance expression. Both gene stacking and gene pyramiding require combining two or more genes into a single plant, and the combined characteristics developed through this process are referred to as stacked traits (Alemu, 2020).

Table 2. Molecular breeding produces resistant plant varieties against major pests

Marker Assisted Selection (MAS) and Marker Assisted Breeding (MAB)	MAS insects' resistance in rice (Venkanna *et al.*, 2018)
Quantitative resistance (QR)	Resistance to leaf hoppers, and brown plant hoppers
Multiple resistance	Multiple insects' resistance in maize (Badji *et al.*, 2018)
Durable resistance	Durable management of insect pests (Sandhu and Kang, 2017)
Gene stacking	Insect resistance in rice (Das and Rao, 2015)
Gene pyramiding	Pyramiding of two insect resistance genes in rice (Liu *et al.*, 2016)

Genetic Engineering for Pest Management

Genetic engineering, also called genetic modification, is the technique of direct, controlled, and artificial modification of an organism's genes that are used to change the genetic makeup of cells, including the transfer of genes within and across species boundaries, to create improved or entirely new organisms.

Therefore, plant genetic engineering is a technology that modifies the genome of a host crop by introducing a foreign donor gene under the control of specific gene regulatory sequences. There are different types of genetic engineering based on the differences between the gene donor and recipient, namely transgenic, intragenic, cisgenic, and subgenics.

i) Transgenics

Transgenics are a term used to describe genetic engineering processes that include transferring foreign genes from unrelated species. This category of genetically modified plants includes the majority of GM products over the last two decades (Sharma *et al.*, 2012).

ii) Intragenics, Cisgenics, and Subgenics

These are several GE techniques that involve gene transfer within the same species or closely related species.

Cisgenesis is the process of genetically modifying a crop species using a complete copy of a natural gene along with its regulatory components, which are only found in sexually compatible plants (same crop species). Intragenesis is the transfer of novel gene and regulatory sequence combinations from one individual of the same species (belonging to the same breedable gene pool) to another species (Holme *et al.*, 2013). The crop that has been genome-edited is referred to as "subgeneics" regardless of whatever genome editing technology is employed (Sticklen, 2015).

3. RNA Interference (RNAi)

RNA interference (RNAi), the sequence-specific suppression of gene expression, provides several opportunities, particularly for the study of gene function, the control of pest populations, and the elimination of disease-causing pathogens. RNAi is used to control insect pests by suppressing essential genes, leading to reduced fitness and mortality.

Target genes are identified in numerous insect pests with the aim of applying RNA interference (RNAi) in pest management. RNAi targets identified include genes involved in feeding and digestion, reproductive fitness, resistance to insecticides and plant allelochemicals, the production of dsRNases, and the transmission of plant viruses (Alemu, 2020).

RNAi in pest management

RNAi protected the plant against a lepidopteran

The first RNAi-based insecticide, SmartStax Pro®, was developed by Monsanto and Dow AgroSciences and approved by the United States Environmental Protection Agency (EPA). This plant-incorporated protectant (PIP) will use a different strategy: this plant will express a variety of Bt proteins as well as dsRNA that targets the WCR Snf7 gene against western corn rootworms (Head *et al.*, 2017). Early investigations on the western corn rootworm, *Diabrotica virgifera virgifera* (WCRW) (Baum *et al.*, 2007), and the cotton bollworm, *Helicoverpa armigera* (CBW) (Mao *et al.*, 2007) provide proof of concept for the use of RNAi in insect pest management. The first successful RNAi plant protected against *H. armigera*was demonstrated by silencing the **CYP6AE14 gene**, necessary for detoxifying **gossypol content from the cotton** crop (*Gossypum hirsutum*). These studies have been conducted to explore alternative target genes to control and understand the RNAi mechanism in lepidopterans. In *H. armigera*-fed transgenic tobacco plants producing dsRNA targeted to the nuclear ecdysone receptor complex (HaEcR), completely necessary for insect development, larval death, and molting abnormalities were found (Mao *et al.*, 2007). A lepidopteran pest, *Spodoptera exigua*, was more easily resistant to the transgenic tobacco that expressed dsRNA for HaEcR.

When Baum *et al.* (2007) exposed larval WCRW to 290 dsRNAs, they found 14 genes that affected larval performance. One of them, vacuolar ATPase subunit A (V-ATPase), was continued for further examination. Low dosages of orally administered V-ATPase-targeting dsRNA in an artificial diet inhibited the corresponding WCRW mRNA. Importantly, larvae raised on maize plants that have been genetically modified to express V-ATPase dsRNA also showed lower expression of the V-ATPase gene and significantly decreased plant root damage. (Baum *et al.*, 2007).

RNAi protected the plant against a coleopteran

According to research, a pest beetle, Henosepilachna vigintioctopunctata, can survive for a shorter period after being exposed to exogenously applied gene-specific dsRNA. They were able to successfully cause 98%, 88%, and 60% mortality in first and third instars and adults after 10, 12, and 14 days by topically applying bacterially expressed dsRNA targeting the Snf7 gene, which encodes an important cellular component of endosomal sorting complexes necessary for transport. Similar to this, Ullah *et al.* (2020) also observed that after exposure to dsRNA targeting the chitin synthase 1 gene, survival and fecundity were considerably decreased in the cotton-melon aphid, *Aphis gossypii*, by 41% and 48%, respectively.

Advantages

- The RNAi pest management method is an increase in our understanding of the developmental, physiological, and molecular mechanisms of insect life stages, but also a great opportunity to develop new methods for insect pest management.

- RNAi is used for large-scale functional genomics studies in many economically important insect pests.
- RNAi pest management techniques are used to identify the target genes for insecticide resistance and also improve our understanding of the mode of action of insecticides and the molecular mechanisms of insecticide resistance.
- RNA Interference is a powerful research tool and budding technology for effective pest management in agriculture.

Limitation

- Lack of reliable dsRNA delivery methods
- The inability to produce large amounts of dsRNA using economical methods.
- It is difficult to achieve efficient RNAi if the trigger is rapidly degraded in an organism unless a high dose of dsRNA or siRNA is applied.
- The easy development of resistance against RNAi products by mutations of target genes, mutations of RNAi core machinery genes, increased dsRNA excretion, enhanced dsRNA degradation, and decreased dsRNA uptake and transport.

4. Genome Editing Technologies

Recent years have seen a rapid development of a variety of new genetic modification techniques (nGMs) or new breeding techniques (NBTs), which are used for precisely modifying the genetic structure of a plant variety to introduce new traits or modify existing ones.A range of new genetic modification techniques (nGMs) or new breeding techniques (NBTs) has been rapidly developed in recent years and used to modify the genetic makeup of a plant variety in a more precise way to introduce new traits or modify existing ones. NBTs include novel methods that are commonly referred to as "genome editing" and that attempt to rapidly create characteristics that may be useful, like in CRISPR-Cas9 genome editing technology (Alemu, 2020).

Genome editing allows genetic material to be added, removed, or altered at particular locations in the genome. A CRISPR array is composed of a series of repeats interspaced by spacer sequences acquired from invading genomes of foreign organisms. CRISPR-Cas9, which stands for clustered regularly interspaced short palindromic repeats and CRISPR-associated protein 9, is a more recent one. The CRISPR-Cas9 technology is more effective, faster less expensive, and more accurate.

Role of CRISPER CAS system in insect application

1. Orcomut ants in *Locusta migratoria*

A miRNA (miR-276) gene was found to facilitate the synchrony of egg development and to directly determine the density of locusts. Insects depend significantly on their olfactory system. Chemical pheromone signals are initially taken in by peripheral tissues, then processed by antennal nerve tissues, before being integrated into the brain by the olfactory and other sensory organs to control insect behaviors including foraging, feeding, and mating. Using a sgRNA that targets the odorant receptor co-receptor (Orco) gene, scientists modified the locust genome using the CRISPR/Cas9 method. The authors successfully generated Orco homozygous and heterozygous mutant lines and achieved highly effective mutation rates in the Orco gene by microinjecting Cas9-mRNA and Orco-sgRNA into locust eggs (Sun*et al.*, 2017). This suggested new ideas for locust control and can be used for other insects from the orthopteran clade such as crickets.

2. CRISPR/Cas9 mediated knockout of the abdominal-A homeotic gene in *Spodoptera furgiperda*

This abdominal A gene is specifically responsible for the formation of abdominal segments. Sfabd-A single guide RNA (sgRNA) and Cas9 protein were injected into 244 FAW embryos in a study by Wu *et al.* (2018). Sixty-two embryos injected with Sfabd-A sgRNA hatched. Twelve of these hatching eggs eventually transformed into larvae that exhibited typical aba-A mutant characteristics, including fused segments. The majority of mutant moths were sterile, and one female produced a few unviable eggs when it was outcrossed to a wild-type male. Genotyping of 20 unhatched Sfabd-A sgRNA-injected embryos and 42 moths that emerged from those embryos revealed that 100% of the unhatched embryos and 50% of the moths had indel mutations at the Sfabd-A genomic locus close to the guide RNA target site (Wu *et al.*, 2018). These findings prove that the CRISPR/Cas9 system is highly effective at editing the genome of the FAW.

3. Eradicating Malaria with CRISPR

By altering the sex-determining gene with CRISPR, researchers have made the male gene predominate over the female gene. After that, the altered organism was spread by gene drive in the expectation that females would eventually be eliminated from the population (Prabhune, 2019). Malaria can only be transmitted by female mosquitoes, so eliminating females would reduce the rate of human transmission.

4. CRISPR Mosquito: Gene Knockout Creates Malaria-Resistant Population

By eliminating a gene, mosquitoes can become resistant to the malaria parasites. Using CRISPR technology, they examined the effects of inactivating the gene for fibrinogen-related protein 1 (FREP1). RNAi-mediated gene silencing techniques had previously been used to study FREP1. The incidence of P. falciparum infection was reduced by 11% as a result of FREP1 silencing, however just as RNAi-based gene silencing only leads to partial protein depletion (Prabhune, 2019). On the other side, CRISPR can be used to completely knock out the gene, and this is used to disrupt the FREP1 gene. Using CRISPR, researchers were able to demonstrate that FREP1 knockout mosquitoes exhibited more slowly than their wild-type counterparts, were less likely to feed blood meals when given the chance, and laid fewer and less viable eggs.

5. Malformations in *Dendrolimus punctatus*

In China and Southeast Asia, the pine caterpillar moth, *D. punctatus*, causes severe defoliation and lowers resin production, making it a devastating forest pest. To precisely and effectively alter gene expression in *D. punctatus*, researchers first introduced mutations into the Wnt-1 gene and indicated that the Wnt-1 gene is associated with segmentation and development. According to the findings, various mutant phenotypes, including deformed heads, legs with defects, and abnormal posterior segments, were seen. Therefore, they may provide a management strategy for a major defoliator (Sun *et al.*, 2017).

6. CRISPR / Cas9 technology in Drosophila

As a model insect, *D. melanogaster* and the application of CRISPR / Cas9 technology in Drosophila has promoted the use of genome editing technology in other insects. CRISPR/Cas9 technology is used to introduce mutations into the Drosophila genome, generating a 4.6 kb deletion in the yellow locus by utilizing two target sgRNAs and a single-stranded oligonucleotide donor (ssODN) template (Singh,2022). It reduced pigmentation and reduced larval mouthparts. Furthermore, it has potential applications for generating designer flies.

References

Alemu, M., (2020). Trend of biotechnology applications in pest management: a review. International Journal of Applied Sciences and Biotechnology, 8(2): 108-131

Badji, A., Otim, M., Machida, L., Odong, T., Kwemoi, D.B., Okii, D., Agbahoungba, S., Mwila, N., Kumi, F., Ibanda, A. and Mugo, S., (2018). Maize combined insect resistance genomic regions and their co-localization with cell wall constituents revealed by tissue-specific QTL meta-analyses. Frontiers in Plant Science, 9: 895

Baum, J.A., Bogaert, T., Clinton, W., Heck, G.R., Feldmann, P., Ilagan, O., Johnson, S., Plaetinck, G., Munyikwa, T., Pleau, M. and Vaughn, T. (2007). Control of coleopteran insect pests through RNA interference. Nature Biotechnology, 25(11): 1322-1326

Chauhan, P. and Kumar, P. (2015). Molecular Markers: Application in Plant Improvement Programmes. International Journal of Applied and Pure Science and Agriculture, 1(07): 74-90

Das, G. and Rao, G.J.N. (2015). Molecular marker assisted gene stacking for biotic and abiotic stress resistance genes in an elite rice cultivar. Frontiers in Plant Science, 6: 698

Drmić, Z., Tóth, M., Lemić, D., Grubišić, D., Pospišil, M. and Bažok, R. (2017). Area wide mass trapping by pheromone based attractants for the control of sugar beet weevil (Bothynoderes punctiventris Germar, Coleoptera: Curculionidae). Pest Management Science, 73(10): 2174-2183

Duhan, J.S., Kumar, R., Kumar, N., Kaur, P., Nehra, K. and Duhan, S. (2017). Nanotechnology: The new perspective in precision agriculture. Biotechnology Reports, 15: 11-23

Gangwar, R.K. (2017). Role of Biological Control Agents in Integrated Pest Management Approaches. Acta Scientific Agriculture 1: 09-11

Gozel, U. and Gozel, C. (2016). Entomopathogenic nematodes in pest management. Integrated pest management (IPM): environmentally sound pest management, 55

Guldmann, E. and Huulgaard, R.D. (2020). Barriers to circular business model innovation: A multiple-case study. Journal of Cleaner Production, 243: 118-160

Head, G.P., Carroll, M.W., Evans, S.P., Rule, D.M., Willse, A.R., Clark, T.L., Storer, N.P., Flannagan, R.D., Samuel, L.W. and Meinke, L.J. (2017). Evaluation of SmartStax and SmartStax PRO maize against western corn rootworm and northern corn rootworm: efficacy and resistance management. Pest Management Science, 73(9): 1883-1899

Haymer, D. (2016). Molecular genetics and insect pest management: Genética molecular y control de plagas de insectos. Revista Agronómica del Noroeste Argentino, 36(1): 9-18

Khera, P., Pandey, M.K. and Varshney, R.K. (2013). Pest and diseases: old and new threats–Modern breeding tools to tailor new crop cultivars. Science et changementsplanétaires/ Sécheresse, 24(4): 261-273

Kirk, W.D. (2017). The aggregation pheromones of thrips (Thysanoptera) and their potential for pest management. International Journal of Tropical Insect Science, 37(2): 41-49

Liu, Y., Chen, L., Liu, Y., Dai, H., He, J., Kang, H., Pan, G., Huang, J., Qiu, Z., Wang, Q. and Hu, J. (2016). Marker assisted pyramiding of two brown planthopper resistance genes, Bph3 and Bph27 (t), into elite rice cultivars. Rice, 9: 1-7

Mao, Y.B., Cai, W.J., Wang, J.W., Hong, G.J., Tao, X.Y., Wang, L.J., Huang, Y.P. and Chen, X.Y. (2007). Silencing a cotton bollworm P450 monooxygenase gene by plant-mediated RNAi impairs larval tolerance of gossypol. Nature Biotechnology, 25(11): 1307-1313

Pickett, J.A., Woodcock, C.M., Midega, C.A. and Khan, Z.R. (2014). Push–pull farming systems. Current Opinion in Biotechnology, 26: 125-132

Prabhune, M. (2019). Eradicating malaria with CRISPR: Mosquito gene editing approach. Synthego, The Bench Blog

Saha, T. and Chandran, N. (2017). Chemical ecology and pest management: A review. Int J Chem Stud, 5(6): 618-21

Saiyad, S.A. (2017). Application of Bacillus thuringiensis as an effective tool for insect pest control. Journal of Agriculture and Veterinary Science, 10(7): 27-29

Sandhu, S. and Kang, M.S. (2017). Advances in breeding for resistance to insects. Breeding insect resistant crops for sustainable agriculture, pp. 67-99

Sharma, H.C., Crouch, J.H., Sharma, K.K., Seetharama, N. and Hash, C.T. (2002). Applications of biotechnology for crop improvement: prospects and constraints. Plant Science, 163(3): 381-395

Singh, N.K. (2018). Reverse breeding: accelerating innovation in plant breeding. Journal of Pharmacognosy and Phytochemistry, 7(1S): 1811-1813.

Singh, S., Rahangdale, S., Pandita, S., Saxena, G., Upadhyay, S.K., Mishra, G. and Verma, P.C. (2022). CRISPR/Cas9 for insect pests management: A comprehensive review of advances and applications. Agriculture, 12(11):1896.

Stenberg, J.A. (2017). A conceptual framework for integrated pest management. Trends in Plant Science, 22(9): 759-769.

Sticklen, M. (2015). Transgenic, cisgenic, intragenic and subgenic crops. Adv Crop Sci Tech, 3(2): e123.

Sultana, R., Kumar, S. and Yanar, D. (2017). Application of entomopathogenic fungi for insect pest's control. Journal of Entomology and Zoology Studies, 5(6): 07-13.

Sun, D., Guo, Z., Liu, Y. and Zhang, Y. (2017). Progress and prospects of CRISPR/Cas systems in insects and other arthropods. Frontiers in Physiology, 8: 608.

Ullah, F., Gul, H., Tariq, K., Desneux, N., Gao, X. and Song, D. (2020). Functional analysis of cytochrome P450 genes linked with acetamiprid resistance in melon aphid, Aphis gossypii. Pesticide Biochemistry and Physiology, 170: 104687.

Venkanna, V., Hari, Y., Rukminidevi, K., Chandra, B.S., Raju, J., Malathi, S. and Reddy, P.R.R. (2018). Markers assisted selection for pyramiding of gallmidge resistance genes in Kavya, a popular rice variety. Int J Curr Microbiol App Sci, 7(4): 745-753.

Wu, K., Shirk, P.D., Taylor, C.E., Furlong, R.B., Shirk, B.D., Pinheiro, D.H. and Siegfried, B.D. (2018). CRISPR/Cas9 mediated knockout of the abdominal-A homeotic gene in fall armyworm moth *(Spodoptera frugiperda)*. PLoS One, 13(12): e0208647.

4

Molecular Methods for The Control of Insect Pests

Ravi Kumar[1*], Vishwajeet Yadav[2], Alamgir[3], Vyankatesh Dhanraj Bagul[3] and Pankaj Singh[4]

[1]Department of Agriculture, Shri Ram College, Muzaffarnagar, Uttar Pradesh, India*

[2]CSIR National Botanical Research Institute, Lucknow, Uttar Pradesh, India

[3]Department of Agricultural Biotechnology, Sardar Vallabhbhai Patel University of Agriculture and Technology, Meerut-250110, Uttar Pradesh, India

[4]Department of Plant pathology, Sardar Vallabhbhai Patel University of Agriculture and Technology, Meerut-250110, Uttar Pradesh, India

---

Introduction

Millions of dollars are lost each year due to insect crop devastation, and the use of chemical pesticides has unwanted side effects such as resistance and unintentional impacts on species that are beneficial (Kumar *et al.* 2019). For farmers, authorities, and academics, finding additional target-selective pest management techniques is of utmost importance. Enhancing performance, cost-effectiveness, and increasing the markets for bio pesticides are all made possible by biotechnology. Molecular methods are used to detect and track the emergence and spread of particular types that are natural enemies (Tipvadee, 2002). It offers chances for the growth of beneficial insect natural enemies with features including pesticide resistance, cold tolerance, and sex ratio change. By using transgenic technology, GE has produced beneficial GM insect-resistant cultivars. Crops that express *Bacillus thuringiensis* (Bt) Cry toxins have been planted all over the world and are an essential tool for pest management. The creation and use of transgenic plants containing insecticidal genes will reduce the need for chemical applications and boost the activity of natural enemies. Molecular methods are used to detect and track the emergence and spread of certain natural enemy biotypes. Biotechnology and genetic engineering can significantly enhance the production, formulation, and storage of entomopathogenic fungi (Yadav *et al.*, 2018). Insect-resistant crops have been created using transgenic technology during the past 20 years to lower yield loss and pesticide use (Christou *et al.*, 2006). Transgenic plants that express genes for the *Bacillus thuringiensis*, or Bt, delta-endotoxin, protease inhibitors, -amylase inhibitors, enzymes, and plant lectins have been shown to be resistant to insects (Padul *et al.*, 2012). One of the most remarkable developments in plant transgenic technology, which has seen tremendous success both commercially and environmentally, is the development of Bt (*Bacillus thuringiensis*) insect-resistant crops.

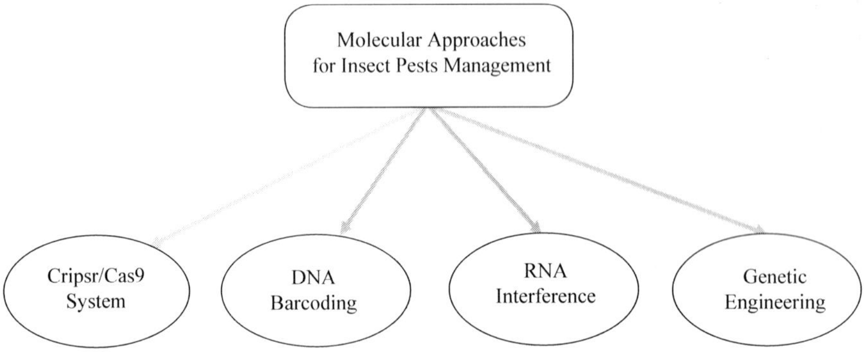

Fig. 1. Different molecular based approaches for insect pest management

A process known as RNAi occurs when cells are exposed to double-stranded RNA (dsRNA), which causes the paired messenger RNA (mRNA) to degrade in a sequence-specific way. This technique has been suggested as a viable method for controlling insect pests since suppressing expression of critical genes might cause insect mortality (Kunte *et al.*, 2020).

Since its discovery as a bacterial adaptive immune system, CRISPR/Cas9 has been modified into a ground-breaking method of editing the genome. CRISPR/Cas9-based editing of genomes has been used to a broad spectrum of species for diverse scientific reasons due to its high degree of specificity and ease of handling. Different methods have been used for pest control using CRISPR/Cas9-like systems, which makes the pests susceptible to different insecticides, impairing the pests' ability to reproduce, preventing the pests from metamorphosing, and providing numerous additional advantages (Singh *et al.*, 2022). The system of CRISPR/Cas9 has the potential to provide novel ways for insect pest management. In this chapter, we covered how to manage insect pests utilising molecular techniques such as DNA barcoding, editing the genome (CRISPR/Cas technology) or GE, and RNAi).

Using CRISPR/Cas technology for pest control

This technology used to create insect-resistant plants seems to be incredibly effective in reducing production costs and boosting farm profitability. The production of plant types that are resistant to insects or other methods for the sustainable control of insect pests to maintain food security show great potential as a result of the fast advancement in CRISPR technology procedures and their many applications. It has recently developed into a tool for genome editing that is very efficient, quick, and quick (Moon *et al.*, 2022). Mega nucleases, zinc finger nucleases (ZFNs), transcription activator-like effector nucleases (TALENs), and clustered regularly interspaced short palindromic repeats (CRISPR)/CRISPR-associated proteins (Cas) are the four main categories of sequence-specific nucleases used in gene editing to date (Gilles and Averof 2014; Baltes and Voytas, 2015). With the use of this technology, epigenetic editing—also known as modifying numerous genes at onceor even a single nucleotide is now possible.

A method for editing genes or genomes that focuses on a specific DNA region, produces a clean cut at the target spot, and renders a gene inactive or substitutes one gene version with another. It was created in 2012 by Jennifer Doudna, Emmanuelle Charpentier, and Fang Zhang. In 2020, the first two of these scientists shared the Chemistry Nobel Prize. The most recent gene editing technology is this. According to Devashish *et al.* (2015), there are three types of CRISPR-Cas9 editing systems: type I, type II, and type III. In type II systems, the Cas9 nuclease needs DNA that fits a single RNA guide (sgRNA). The guide RNA and the Cas protein are the two main elements of the CRISPR-Cas system. The nuclease enzyme Cas9 protein, also known as the molecular scissors, is in charge of cutting the DNA. Cas9 is guided to the desired location in the genome by molecules called guide-RNAs, where it will sever the old sequence and insert the new one (Courtier-Orgogozo *et al.*, 2015). Figure 1 depicts three distinct CRISPR-based genome-editing techniques for controlling insect infestations.

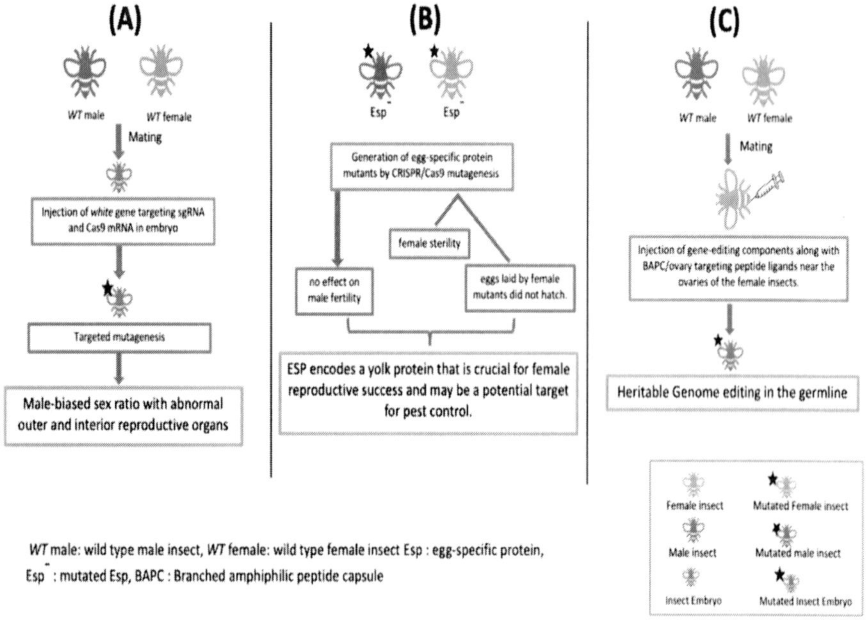

Fig. 2. Various CRISPR-based genome-editing techniques for controlling insect pests. **(A)** Sterilised insect method (SIT) in picture form **(B)**CRISPR/Cas9 targeting of female reproductive wellness**(C)**Ovary targeting signals are injected into CRISPR/Cas9 components to provide permanent genome modification in the germline.

Source: https://doi.org/10.3390/ agriculture12111896

Recently, a theory on the application of gene editing methods for plant pest management emerged. With the use of CRISPR/Cas9 technology, Zuo*et al.* (2017) altered *Spodoptera exigua*, producing a mutant insect with a high level of resistance to the insecticides chlorantraniliprole, cyantraniliprole, and fubendiamide. *Helicoverpa armigera* was modified using CRISPR/Cas9 to alter the cadherin gene in order to show how the gene contributes to the development of resistance against Bt toxin. Similar to other significant agricultural pests, the beetroot armyworm (*Spodoptera exigua*) is generally managed using chemical pesticides. Resistance to the pesticide Spinosad has recently been identified in S. exigua field populations. Zuo *et al.* (2020) effectively created a homozygous strain (Se6-KO) with a 1760-bp deletion inside Se6 in *S. exigua*, resulting in a premature termination of Se6 via the CRISPR/Cas9 genome-editing method. They convincingly confirmed the functional involvement of Se6 in spinosyn pesticide resistance and offer an example of genome editing being used to validate an objective premature truncation linked with resistance. Moon *et al.* (2022) have examined and explored the application of CRISPR-Cas technology for genome editing in for a long-time insect pest management.

Gene drive, which is based on the CRISPR/Cas-9 gene editing system, is a strong technique which encourages the inheritance of the gene drive tool directly through sexual reproduction, allowing it to spread swiftly through a population. Courtier Orgogozo *et al.* (2017) presented another potential use of CRISPR-based gene drive, notably pest species management to boost crop yield. Singh *et al.* (2022) emphasised recent advances in CRISPR/Cas9 investigation and explored its potential applications in pest control.

DNA barcoding for pest control

The successful implementation of an integrated pest management (IPM) programme depends on accurate identification of agricultural pests. However, additional molecular approaches like DNA barcoding are utilised since conventional morphological methods have certain limits. Especially for juvenile stages when morphometric investigations are confusing, DNA barcoding is a trustworthy method for identifying agricultural pests. It will also aid in the creation of more potent pest management strategies for controlling pest species. In order to assess the precision of phenotypic identification of insect pests through DNA barcoding, Tahir *et al.* A total of 247 insect pests comprising 18 genera, 10 families, and 22 species were recognised morphologically. Nevertheless, molecular identifications proved the existence of 20 species, 11 families, and 16 genera of agricultural pests. Using this molecular method, specimens that were misclassified through morphological investigations were assigned to the correct taxon. Results showed that some pest species have apparent barcode gaps.

There are two pest species called *Tribolium confusum* and *Tribolium castaneum* (Coleoptera: Tenebrionidae) that are extremely difficult to distinguish from one another based just on appearance. Such pests are identified in grain that has been kept in their juvenile stages, which makes identification more challenging. Integrative pest control urgently needs precise identification of these pests. DNA barcoding was utilised by Aslam *et al.* (2019) to precisely identify these pests at any stage of life. To enhance invasive pest detection at U.S. ports of entry in 2019, Madden *et al.* employed DNA barcoding.

Thrips are a main sucking pest and plant virus vector that causes significant economic damage in agriculture. Thrips identification is critical for efficient pest management tactics. Tyagi *et al.* (2017) undertook the greatest DNA barcoding effort on 370 sequences from 89 thrips morphospecies, comprising 104 unique sequences from 39 morphospecies, include four species' type specimens. For 73 species (82%), the results of various species delimitation approaches (BIN, ABGD, GMYC, and bPTP) were compatible with the corresponding morphological identifications. The presence of more than one MOTU in 14 morphospecies suggests cryptic diversity, which includes two significant vector species (*Frankliniellas chultzei* and Thrips palmi). Yet, four morphospecies (*Thrips moundi, Thrips carthami, Haplothrips andersi*, and *Haplothrips gowdeyi*) had low genetic distances with overlapping in the barcode gap, requiring further analysis with multiple molecular markers along with additional specimens from diverse geographical areas for better taxonomic judgement.

RNA interference (RNAi)

It is a new technique that appears to be a viable alternative to chemical pesticides. RNAi allows for the silencing of critical genes in insects, resulting in the selective destruction of specific species. Dicer (a member of the ribonuclease III family of nucleases) first converts dsRNA into 20-25bp pieces known as small interfering RNA (siRNA). Following that, siRNAs are taken up by the RNA induced silencing complex (RISC), where they are unravelled into two single strands, the "guide" strand and the "passenger" strand. The complex's "passenger" strand is removed, and the "guide" strand subsequently guides the RISC complex to connect with the complementary mRNA, inhibiting translation of this target mRNA." The RISC component argonaute-2 protein is crucial in the identification and mRNA degradation process. It has the PAZ and PIWI domains. The guide strand, which acts as a template to trap the desired mRNA, is kept by the PAZ domain. According to Yu *et al.* (2013) and Darrington *et al.* (2017), the PIWI domain splits the trapped mRNA into two pieces that are then further broken down by cytoplasmic nucleases.

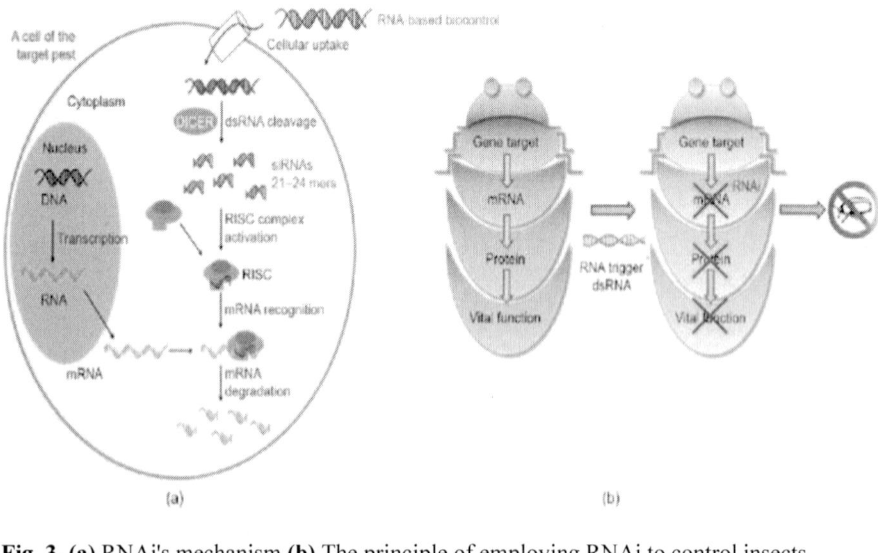

Fig. 3. (a) RNAi's mechanism **(b)** The principle of employing RNAi to control insects.

Source: http://dx.doi.org/10.1016/j.eng.2019.09.008

Insects are exposed to dsRNA via three major routes: eating, topical application, and injections into the hemolymph. In 2020, Kunte *et al.* examined the administration of dsRNA by food. In conclusion, new techniques have been established to increase the effectiveness of this approach, including the use of transgenic plants, microbes, and formulations based on nanotechnology. The challenges that must be solved for this technology to be effective as a pest management tool were highlighted by Kunte *et al.* in 2020. Similar to this, Adeyinka *et al.* (2020) highlighted the existence of nucleases and double-stranded RNA binding protein while describing certain genes implicated in insect dsRNA absorption and discussing the causes of variable RNAi response in insect pests. Recent innovations in dsRNA delivery for effective and efficient knockdown of insect pests were examined. In conclusion, topical dsRNA administration in conjunction with a complex nanoparticle has enormous promise for RNAi-mediated pest management. In their study of RNAi-mediated control of hemipteran insects, Jain *et al.* (2021) offered viable solutions, such as in-silico strategies combined with laboratory-based toxicity testing to avoid any unintended impacts on helpful species. The researchers subsequently investigated strategies to reduce dsRNA degradation in the environment and in insects, such as stacking and dsRNA effector formation. Finally, to minimise the cost of dsRNA manufacture and application and to enhance broad-spectrum hemipteran pest control, we investigate non-transformative RNAi techniques, concatomerization of RNAi sequences, and pyramiding RNAi with active components.

Horak (2020) sequence-directed suppression of protein synthesis by RNA interference has recently been proposed as a viable method for controlling pest species. RNAi, which is intended to be species-specific, has a lower impact on non-target species and the environment. It is important to conduct further study to advance the field of RNAi-based control of vertebrate pest species. According to Yan *et al.* (2021), nanoparticles like chitosan, liposomes, and cationic dendrimers have advantages over traditional double-stranded (ds)RNA delivery systems for delivering dsRNA/small interfering (si)RNA to increase RNAi efficiency. This advances the research and application of RNAi-based pest management strategies.

The challenges of traditional dsRNA delivery systems are illustrated, the mechanism of nanoparticle-mediated RNAi is revealed, recent developments and effective uses for nanoparticle-mediated RNAi in pest management are summarised, and finally the potential of nanoparticle-based RNA pesticides are addressed. The use of RNAi encourages the development of innovative ways to plant protection in an environmentally friendly way. Commercially viable genetically engineered crops expressing dsRNA have been produced as insect pest control tools. List *et al.* (2022) discovered interesting trends that support the assumption that fundamental toxicological principles applicable to RNAi bioassays and give preliminary guidance for future study comparable to studies on conventional insecticides. To speed the development of RNAi as an effective pest management tool, we urge for training that incorporates genetic, organismal, and toxicological techniques.

Genetic engineering (GE)

Although the word "genetic engineering" is frequently perceived as emotional or even frivolous, it is likely the one that the majority would choose. The techniques used to manipulate genes might be referred to by a variety of names. recognise. However, a number of alternative terminologies, such as genetic manipulation, cloning of genes, recombinant DNA technology, and the new genetics, can also be used to characterise the technique (Nicholl, 2023). A multifaceted integrated pest management strategy should include genetically modified crops that offer resistance to pests and illnesses or tolerance to herbicides. Unfortunately, the majority of these methods use natural resources, but genetic engineering has lately become a practical substitute for controlling plant insect pests. When introduced into bacteria using r-DNA technology, insecticidal proteins have demonstrated effectiveness against coleopteran and lepidopteran pests. Native expression of secondary plant metabolites, such as flavonoids, exhibits detrimental effects on the development of insect pests.

Insecticidal potential of plants engineered to produce enzymes that break down insect cell walls, which include structural polysaccharides like chitin, was effective. The fundamental idea behind food digestion is the hydrolysis of the meal by an enzyme. When insects encounter enzymatic inhibitors (which include protease inhibitors, alpha-amylase inhibitors, etc.) found in the feeding section of the plants, such gut enzymes become attenuated. These genes, which all produce insecticides, are widely distributed in the natural world (Dash *et al.* (2021). The development of transgenes against plant insect pests via genetic engineering is a triumph in this field.

Conclusion

Pest insects pose a significant risk to agricultural production globally, contributing significantly to economic loss and endangering global food security. Numerous pests have evolved a strong resistance to chemical insecticides as a result of their widespread use, and some of these pesticides have unintended side effects on pollinators and natural foes. The development of insect pest resistant plants using the DNA barcoding gene editing, genetic engineering and RNAi looks to be a very effective way to reduce production costs and boost farm profitability. For ecologically friendly practises, the advancement of molecular crop protection strategies is necessary.

References

Adeyinka, O.S., Riaz, S., Toufiq, N., Yousaf, I., Bhatti, M.U., Batcho, A. Tabassum, B. (2020). Advances in exogenous RNA delivery techniques for RNAi-mediated pest control. Molecular Biology Reports, 47: 6309-6319

Aslam, A.F.M., Sultana, S., Das, S.R. and Howlader, A.J. (2019). DNA barcoding to resolve the confusion in identifying Tribolium confusum and Tribolium castaneum. Bangladesh J. Zool, 47(2): 333-342

Baltes, N.J., Voytas, D.F. (2015). Enabling plant synthetic biology through genome engineering. Trends Biotechnol, 33: 120-131

Christou, P., Capell, T., Kohli, A., Gatehouse, J.A., Gatehouse, A.M. (2006). Recent developments and future prospects in insect pest control in transgenic crops. Trends in Plant Science.11: 302-308

Courtier Orgogozo, V., Morizot, B. and Boëte, C. (2017). Agricultural pest control with CRISPR based gene drive: time for public debate: should we use gene drive for pest control? EMBO reports, 18(6): 878-880

Darrington M., Dalmay T., Morrison N.I., and Chapman T. (2017). Implementing the sterile insect technique with RNA interference – a review. Entomol Exp Appl 164: 155–175

Dash, L., Rout, S., Mishra, U., Sahoo, G., Prusty, A.K. (2021). Insecticidal genes in Pest Management. Annals of the Romanian Society for Cell Biology, 25(6): 5601-5608

Devashish, R., Lina, A., Archana, R., Magnus, L. (2015). The CRISPR-Cas immune system: Biology, mechanisms and applications. Biochimie, 117: 119-128

Doudna, J.A., Charpentier, E. (2014). The new frontier of genome engineering with CRISPR-Cas9. Science, 346: 1258096

Gilles, A.F., Averof, M. (2014). Functional genetics for all: Engineered nucleases, CRISPR and the gene editing revolution. EvoDevo, 5: 43

Horak, K. (2020). "RNAi: Applications in Vertebrate Pest Management" (2020). USDA National Wildlife Research Center - Staff Publications. 2380. https://digitalcommons.unl.edu/icwdm_usdanwrc/2380

Jain, R.G., Robinson, K.E., Asgari, S., Mitter, N. (2021). Current scenario of RNAi based hemipteran control. Pest Management Science, 77(5): 2188-2196

Kumar, S., Nehra, M., Dilbaghi, N., Marrazza, G., Hassan, A.A., Kim, K.H. (2019). Nano-based smart pesticide formulations: Emerging opportunities for agriculture. J Control Release 294: 131-153

Kunte, N., McGraw, E., Bell, S., Held, D., Avila, L.A. (2020). Prospects, challenges and current status of RNAi through insect feeding. Pest Management Science, 76(1): 26-41

List, F., Tarone, A.M., Zhu Salzman, K., Vargo, E.L. (2022). RNA meets toxicology: efficacy indicators from the experimental design of RNAi studies for insect pest management. Pest Management Science, 78(8): 3215-3225

Madden, M.J., Young, R.G., Brown, J.W., Miller, S.E., Frewin, A.J. Hanner, R.H. (2019). Using DNA barcoding to improve invasive pest identification at US ports-of-entry. PLoS One, 14(9): e0222291

Moon, T.T., Maliha, I.J., Khan, A.A.M., Chakraborty, M., Uddin, M.S., Amin, M.R. Islam, T. (2015). CRISPR-Cas Genome Editing for Insect Pest Stress Management in Crop Plants. Stresses, 2: 493–514. https://doi.org/10.3390/ stresses2040034

Nicholl, D.S. (2023). An introduction to genetic engineering. Cambridge University Press

Padul, M.V., Chougale, A.D., Dama, L.B., Gadge, P.P, Shaikh, F.K., Chopde, R.M., Vadane, S.S. (2012). α-amylase from sugarcane wooly aphid (Ceratovacunalanigera). Treands in Biotechnology Research Journal, 1(1): 49-52

Singh, S.,Rahangdale, S., Pandita, S., Saxena, G., Upadhyay, S.K., Mishra, G., Verma, P.C. (2022). CRISPR/Cas9 for Insect Pests Management: A Comprehensive Review of Advances and Applications. Agriculture, 12, 1896. https://doi.org/10.3390/ agriculture12111896

Tahir, H.M., Noor, A., Mehmood, S., Sherawat, S.M. Qazi, M.A. (2018). Evaluating the accuracy of morphological identification of insect pests of rice crops using DNA barcoding, Mitochondrial DNA Part B 3(2): 1220-1224

Tipvadee, A. (2002). Biotechnology for insect control. Proc. Sat. Forum, 2002, 74-85

Tyagi, K., Kumar, V., Singha, D., Chandra, K., Laskar, B.A., Kundu, S. and Chatterjee, S. (2017). DNA Barcoding studies on Thrips in India: Cryptic species and Species complexes. Scientific Reports, 7(1): 1-14

Yadav, T., Singh, V., Acharya, V. S., Yadav, R. (2018). Molecular approaches and their role in integrated pest management, Journal of Pharmacognosy and Phytochemistry, 7(6): 145-150

Yan, S. Ren, B.Y., Shen, J. (2021). Nanoparticle mediated double stranded RNA delivery system: A promising approach for sustainable pest management. Insect Science, 28(1): 21-34

Yu N, Christiaens O, Liu J, Niu J, Cappelle K, Caccia S, et al. (2013). Delivery of dsRNA for RNAi in insects: An overview and future directions. Insect Sci 20: 4–14

Zuo Y., Wang H., Xu Y. et al (2017). CRISPR/Cas9-mediated G4946E substitution in the ryanodine receptor of Spodoptera exigua confers high levels of resistance to diamide insecticides. Insect Biochem Mol Biol 89: 79–85

Zuo, Y., Xue, Y., Lu, W., Ma, H., Chen, M., Wu, Y., Hu, Z. (2020). Functional validation of nicotinic acetylcholine receptor (nAChR) α6 as a target of spinosyns in Spodoptera exigua utilizing the CRISPR/Cas9 system. Pest Management Science, 76(7): 2415-2422

5

Role of Bio-Inputs in Pest Management

Madan Mohan Bajpeyi[1*], Navaneet Kumar[2], Anukiran Sahu[3]
A.N. Warghat[4] and Effat Jehan[5]

[1*]*Department of Entomology, Chandra Shekhar Azad University of Agriculture & Technology, Kanpur, Uttar Pradesh, India*
[2]*Department of Entomology, Acharya Narendra Deva University of Agriculture and Technology Ayodhya, Uttar Pradesh, India*
[3]*Department of Nematology, Odisha University of Agriculture and Technology, Bhubaneswar, Odisha, India*
[4]*Department of Entomology, Naini Agricultural Institute, Sam Higginbottom University of Agriculture Technology and Sciences, Prayagraj, Uttar Pradesh, India*
[5]*Department of Zoology, University of Kashmir, and Kashmir, India*

Introduction

Biopesticides are biological substances or plant-based products that are used for decreasing the number of pests in an ecosystem. These substances, which include natural enemies, microbial products, phytochemicals, and semiochemicals, are derived from plants and animals. Biopesticides use non-toxic approaches to keep pests under control. They are widely recognised as low-risk items that are secure for both humans and the environment. It makes use of natural pest such as parasitoids, predators, nematodes, and semiochemicals as well as botanicals, microbiological pathogens like fungi, bacteria, and viruses. Synthetic pesticides have created resistance, resurgence and an emergence of secondary pests as an outcome of their unchecked use. Strict restrictions on the use of synthetic pesticides and the increased demand for organic practises have promoted the use of biopesticides. By employing biopesticides, pests can be controlled without endangering the environment or people's health. Biopesticides play a significant role in promoting sustainable agriculture. Prior to reaching the economic threshold level, they significantly reduce arthropod populations, which results in mortality (Betz *et al.*, 2000). Integrated pest management programmes (IPM) urgently need to incorporate biopesticides since they have low phytotoxicity, leave no residues, and are environmentally beneficial. (Bhattacharyya *et al.,* 2016).

Perspectives on Biological Pesticides: Indian and Global Context

The worldwide market for insecticides as a whole is made up of 15% by the bio-pesticides industry, which has a $3 billion global market share and accounts for 5% of all crop protection. At the moment, biopesticides are expanding at a 15% yearly pace. Only 15% of bio inputs for managing insect pests have been found in effective natural control schemes. Over 125 different kinds of natural enemies are marketed as biopesticides. Surprisingly, 90% of the microbial biopesticides currently on the market are made from the single entomopathogenic bacteria *Bacillus thuringiensis* (Bt.). The world's three largest manufacturers and consumers of biopesticides are the United States, the European Union, and India. The India Biological pesticides Market Insights research predicts that the Indian biopesticides market would increase from INR 197 billion in 2018 to INR 316 billion by 2024. Given that pest-related crop losses in India total about USD 9259 million annually, the market for biopesticides has a large potential. However, biopesticides now only account for 5% of the worldwide pesticide business. 200 entomopathogenic fungi-based products are already available, and the Central Insecticides Board and Registration Committee (CIBRC) of India has registered around 970 microbiological formulations. Beauveria bassiana, *B. brongniartii, Metarhizium anisopliae, Lecanicillium lecanii*, and *Hirsutella thompsonii* are frequently used to manage arthropods. Additionally, there are 45 commercially available Bt-based products, with the kurstaki strain of *B. thuringiensis* being heavily employed against lepidopteran pests like loopers and bollworms. The *B. thuringiensis israelensis* and *sphaericus subspecies* are both effective against mosquitoes. Nuclear polyhedrosis viruses (NPV) control both armyworms and bollworms. In India, lepidopteran pests of rice, sugarcane, and apples were successfully controlled for the first time with the aid of *Steinernema carpocapsae* (DD-136 strain). Other natural strains of entomopathogenic nematodes, such as *Heterorhabditis indica, Steinernema carpocapsae, S. abbasi, and S. thermophilum*, as well as S. riobrave, have been used to control pests in agricultural fields. A commercial product with a 12-month shelf life has been created in India using a wettable formulation derived from H. indica.

Classifications of biopesticides

Biopesticides are divided into three main categories by the Environmental Protection Agency based on the type of component they include.

1. Microbial biopsticide.
2. Biochemical pesticide.
3. Plant-incorporated protectants

This category includes the bulk of broad-spectrum bio-pesticides for pest control, also referred to as microbial pesticides. To control insect pests, they use microorganisms including bacteria, viruses, fungus, and protozoans as active components. These biopesticides are highly precise in their targeting of the designated species. Microbial biopesticides exhibit host-specificity, are self-sustaining, and are environmentally safe. Because it effectively eliminates a wide variety of pests, including coleopterans, lepidopterans, and dipterans, Bt. (*Bacillus thuringiensis*) is one of the most widely employed bacteria for insect pest control. Plant kinds that have been genetically modified to produce pesticides are referred to as genetically modified crops, or Plant-Incorporated Protectants (PIPs). The plants are given the genetic material needed to produce these chemicals, which gives them pest resistance. The incorporation of pesticidal proteins derived from bacteria or fungi into the plants gives them the ability to fend off particular pests. An illustrative example of this is the use of *Bt.* protein, which is introduced into plants via genetic engineering to develop PIPs. *Bt.* toxin is highly specific to its host and typically causes rapid mortality in pests, usually within 48 hours PIPs do not affect vertebrates or the environment in any negative ways.

Biochemical pesticides are naturally occurring compounds and secondary metabolites that have the ability to control pests. Herbal pesticides are another name for these substances. In India, neem-based formulations are frequently used as biochemical pesticides. Examples include neem oil, NSKE, and concentrations made from neem bark and leaves. Additional biochemical pesticides include pyrethrin from *Chrysanthemum cinerariaefolium* and essential oils from canola, tea tree, lemongrass (*Cymbopogon citratus*), and tea tree. Diatomaceous earth (DE), which is formed from fossilised sediments of various siliceous species, mostly diatoms and other algae, is used to treat a range of field pests. DE has a high capacity for absorption, which can erode and desiccate the insect's cuticle, killing it instantly. Semiochemicals, which are categorised as biochemical pesticides, include a wide range of compounds, including pheromones used for sex, repellents for insects, and kairomones. These chemical cues, which are mostly produced by insects, cause individuals of the same or different species to change their behaviour. Insect sex pheromones are among the commercially accessible semiochemicals that are frequently used. Through mating disruption techniques, sex pheromones have proven to be very effective in regulating adults. For the management of flying insects, particularly lepidopterans and coleopterans, mass trapping using sex pheromones is very helpful in integrated pest management (IPM) programmes.

Bacterial-Derived Biopesticides

In general, there are three types of bacterial pesticides:

1. **Obligate pathogens:** This pathogen requires certain environments and mediums for sporulation, reproduction, and growth. They only infect a small number of hosts, and many of those hosts can produce spores. These characteristics make them well-suited for implementation in biological pest control programs. Obligate pathogens exhibit good stability for example; populations of white grubs are affected by milky disease caused by Bacillus papillae and *B. lentimorbus*.
2. **Facultative pathogens:** Facultative pathogens do not need particular environment to develop, reproduce and sporulate Compared to obligatory pathogens, they are often less virulent.

   Facultative pathogens can be categorized into two groups:

   i. Crystalliferous

   ii. Non-crystalliferous

Crystal-producing bacteria create proteinaceous crystals that are poisonous. Most of them are capable of forming spores and possess a broad range of hosts. One commercially exploited example of a crystalliferous spore-forming bacterium is *B. thuringiensis berliner*, used as a biocontrol agent.

3. Potential pathogens: Potential pathogens are easily cultivated on synthetic medium and have a wide host range. Insects are effective against a variety of hosts and can become infected even at low levels (10,000 cells or fewer). For instance, Pseudomonas aeruginosa and Serratia marcescens are used to combat root-knot nematodes, while Phyllophaga blanchardi is used to manage coleopteran pests. However, due to their non-invasive and spore-free nature, these pathogens have limited promise as efficient biocontrol agents.
4. Microbial biopesticides have been effectively produced using Saccharopolyspora spinosa, Serratia entomophila, *Bacillus popilliae, B. thuringiensis, Clostridium bifermentans, Pseudomonas alcaligenes, Pseudomonas aureofaciens, and Streptomyces avermitilis.* Lepidopteran, dipteran, coleopteran, and hymenopteran pests are at risk from *B. thuringiensis and Bacillus sphaericus.* The creation of certain crystal proteins by the bacteria during the spore formation phase is principally responsible for the organism's insecticidal characteristics. The targeted pests are poisoned by these proteins. Table 1 provides a thorough list of *B. thuringiensis* subspecies and the associated toxins used to combat different pests.1 uses the toxin produced by the *B. thuringiensis* subspecies, which is effective against many pests. The commercially available *B. thuringiensis*-

based insecticides for controlling different pests are shown in Table 1.2 (Gelernter and Schwab 1993).

Table 1.1: List of *Bacillus thuringiensis* subsp. and its toxin used against various pests

Bacillus thuringiensis subsp.	**Toxin class**	**Target insect**
Bt subsp. *Kurstaki*	Cry I	Lepidoptera
Bt subsp. *aizawai*	Cry I	Lepidoptera
Bt subsp. *Berliner*	Cry I	Lepidoptera
Bt subsp. *kurstaki*	Cry II	Lepidoptera, Diptera
Bt subsp. *tenebrionis*	Cry III	Coleoptera
Bt subsp. *israelensis*	Cry IV	Diptera

Table 1.2: List of insecticides based on Bt. that are sold commercially control to various pest

Bacterial species used	**Target pests**	**Trade name**
Bacillus popilliae	grubs of Japanese beetle	Doom, Milky Spore Disease, Grub Attack
B. thuringiensis subsp. Aizawai	Lepidopterans	Certan
B. thuringiensis subsp. israelensis	Mosquito, blackflies and fungus gnats	Bacticide, Summit Bactimos, Bacto Power-Bti, Biodart M, Larvect50, Deltafix, Biovectra Spicbiob, VectoBac, Vectocid
B. thuringiensis subsp. Kurstak	Lepidopteran larvae	Abtec Btk, Agni, Bioasp, BioDart, Bactur, Biobit, Bioworm, Biolep, B.T.Killer, Caterpilin, Cezar etc.
B. thuringiensis subsp. Galleriae	Lepidopteran larvae, bollworms, Diamondback moth	Spicturin
B. sphaericus	Mosquito larvae	Sphericide, VectoLex
B. thuringiensis subsp. tenebrionis	Colorado potato beetle, stored pests, coleopteran adults and larvae	Foil, M-one, M-track, Novardo, Trident

Mode of Action of Bacillus thuringiensis

When *B. thuringiensis* undergoes spore development, it creates protein crystals that are essential for breaking down gut cells in insects that are sensitive to it. The bacterium produces cry and cyt proteins, which are crystalline proteins. The alkaline pH environment of the insect's stomach, which ranges from 9.0 to 11.0, is activated when the larva consumes the cry protein (Bravo *et al.* 2007). These proteins bind to the receptor location in the gut, causing the cells of the midgut to develop holes. The midgut is consequently paralysed by the Bt.

toxin, which also induces cell lysing and the discharge of gut contents into the insect's hemocoel. The death of the organism is the result of this disruption of the pH equilibrium.

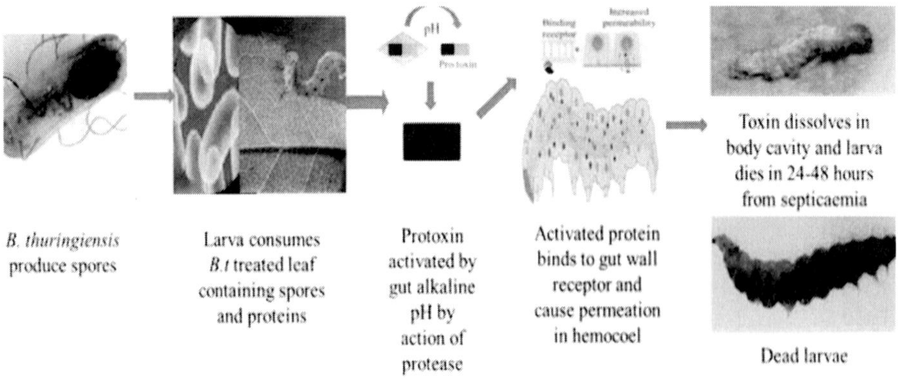

Fig. 1. Mode of action of bacterial pesticide (B. t.)

The benefits of bacterial pesticides

Use of bacterial biopesticides is harmless and does not harm humans, animals, or other living things.

Specificity: Bacterial biopesticides only affect certain types of pests while leaving harmless insects alone.

Compatibility with other techniques: Synthetic chemical biopesticides and the majority of bacterial insecticides can coexist without conflict.

Bacterial biopesticides are non-hazardous and can be used even when the crop is being harvested because they don't leave any residues behind.

Natural persistence: Bacterial biopesticides can establish themselves and maintain their efficacy throughout several crop growing seasons.

Drawbacks of Microbial Insecticides

1. Only particular species or groups of insects are harmed by microbial pesticides. They can successfully eradicate some pests while permitting others to live and possibly cause harm.
2. Synthetic pesticides may harm beneficial creatures when used in pest management programmes that make use of predators and parasitoids. As a result, it is essential to carefully choose the right chemical pesticides.
3. Several conditions, like as heat and ultraviolet rays, can affect how effective microbial pesticides are. As a result, it is suggested that you only use them in the morning.

4. Specific preparation and storage techniques are needed for microbial pesticides. To maintain the infections' virulence, it is essential to store microbial pesticides in accordance with the directions on the package.
5. The challenge of mass-multiplying natural competitors year-round restricts their marketability. Additionally, compared to chemical pesticides, the product registration process and a lack of availability restrict the use of biopesticides.
6. The effectiveness of biopesticides in the field is influenced by a variety of climatic conditions, including temperature, relative humidity, and rainfall. The weather needs to be ideal for biocontrol agents to grow and survive in the field.

Role of viruses as biopesticides

Since baculoviruses (BV) are highly pathogenic to insects and other arthropods, they are frequently utilised as viral biopesticides. BVs are little viruses with double-stranded DNA that spread by parasitic reproduction. There are three subgroups of the genus Baculoviruses: cytoplasmic polyhedrosis viruses (CPVs), granulosis viruses (GVs), and nuclear polyhedrosis viruses (NPVs). These viruses fall under the inclusion virus category. The number of occlusion bodies they possess as well as the pattern of the protein coating around them can be used to identify them. In contrast to NPV and CPV, which make polyhedral bodies made up of several virus particles, GV produces granular occlusion bodies with a single viral particle. Different baculoviruses uncoil NPV, CPV, and GV in different ways. NPVs uncoat within the nucleus, CPVs uncoat in the cytoplasm, and GVs uncoat within the nuclear pore complex to infect the host. The occlusion bodies are essential to the virus' ability to survive outside of the host.

Mode of Action of Viruses

In terms of how they work, viruses operate similarly to bacteria in that they need to be consumed by insect larvae to cause illness. Once swallowed, the virus crosses the intestinal epithelium to enter the insect's body, where it causes systemic infection in the host cells. When it comes to the NPV virus, it passes through the intestinal epithelium and enters the hemocoel, fat body, and other tissue cells. It alters tissue integrity and ultimately causes the cadavers to liquefy. However, the infection brought on by GV is only present in the midgut of the insect (Fig.1.2).

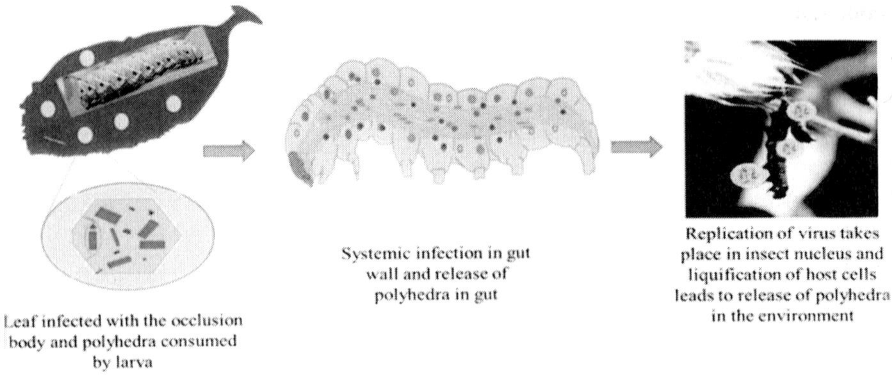

Fig. 2. Mode of action of virus

The Method for Preparing NPV and CPV

Third- to fourth-instar Helicoverpa armigera larvae typically contract viruses by eating infected food. Over a period of 5 to 10 days, the viral disease's last stage manifests itself. The larvae ascend higher in the plant canopy before dying once the virus has completely infected them. Thousands of millions of polyhedra are now ejected by the infected larvae, assisting in the propagation of virus particles from the deceased animals to the canopy's underside. Through infecting healthy larvae, this behaviour promotes in the virus's spread.

In commercial production, larvae are either collected from the field or reared in a lab and fed a contaminated diet. The larvae are deprived for 24 hours prior to virus inoculation. The virus infects the larvae, which are suspended upside-down in tubes, in three to four days. By being soaked in water, infected larvae are collected and disinfected. The cleaned larvae are then crushed in a crusher and pestle with 1:2.5 (w/v) sterile cooled distilled water. The homogenate is repeatedly rinsed with sterile distilled water in a ratio of 1:7.5 to 12.5 (w/v) based on the original weight of the cadaver larvae before being filtered through a double layer of muslin fabric. To separate the polyhedra from the waste, the filtrate is centrifuged. For Helicoverpa armigera larvae, the effective (LC50) of NPV is 1 x 109/ml. HNPV should be used to gramme, pigeon pea, and cotton at 250-300 LE, 500 LE, and 250 LE/ha, respectively. Viruses have a high level of host specificity and can significantly reduce host populations. Commercially available viral biopesticides include *Cydia pomonella granulovirus* (CpGV), *Spodoptera exigua* multi-enveloped nuclear polyhedrosis virus (SeMNPV), and Helicoverpa zea single-enveloped nuclear polyhedrosis virus (HzSNVP). The efficacy of these biopesticides in pest management techniques has been demonstrated.

Table 1.3: list of viral biopesticides and their trade name

Target pest	Host	Products available
NPV of *H. armigera (HaNPV)*	*Helicoverpa zea H. armigera*	Biokill-H, BioVirus-H, Heli-Cide, Heliokill, Helimar-NPV, Helivax, Jas Viro–H, Helicop, Heligard, Somstar-Ha
NPV of *S. litura (SlNPV)*	*Spodoptera litura S. exigua*	BioVirus-S, Jas Viro–S, Spodo-Cide, Spodopterin, Somsta-SL
Pine sawfly NPV	*Larvae of pine sawfly*	Neochek-S
Granulosis virus for Codling moth (GV)	*Codling moth caterpillars*	Madex, Carpovirusine, CYD-X
Nuclear polyhedrosis for Gypsy moth	*Gypsy moth caterpillars*	Gypchek virus
Tussock moth NPV	*Tussock moth caterpillars*	TM Biocontrol-1

Benefits of Viral Biopesticides

1. Viral biopesticides do not affect people or other non-target animals or plants.
2. Viral biopesticides do not cause target organisms to become resistant.
3. Viruses can work in harmony with chemical pesticides as well as other techniques of controlling insects.
4. Viruses self-replicate and efficiently control pest population growth.
5. Secondary pest outbreaks are not brought on by viral biopesticides.
6. Viral biopesticides do not leave behind pesticide residues.
7. Pre-harvest intervals are not necessary for viral biopesticides.

Disadvantages of Viral Biopesticides

1. It can be used to deal with a range of pests, comparable to chemical pesticides.
2. A prolonged period of deadly infection is required for the target pests to be effectively controlled.
3. Viral inactivation can be caused by environmental conditions including ultraviolet radiation, extremely high temperatures, and others.

Fungi as Biopesticides

Due of the wide variety of hosts they can be found on in both field and storage settings, fungi-based biopesticides are extremely adaptable. The use of fungus-based biopesticides is widely recognised as safe and highly targeted. These goods come from parasitic or saprophytic fungi that thrive inside an insect's body and eat the interior tissues. Beauveria bassiana, Metarhizium anisopliae, *Nomuraea rileyi, Trichoderma viride, Paecilomyces farinosus, and Verticillium lecanii* are a few of the most widely used commercial fungal biopesticides on the market. These fungus-based biopesticides have proven successful at reducing pest populations. To maintain their own life, these fungi form either necessary or facultative relationships—symbiotic or antagonistic—with insects. Fungi must come into contact with the intended pests and then infect them in order to have a harmful effect. Aphids, beetles and their larvae, leaf and plant hoppers, pod borers, cutworms, grasshoppers, as well as other lepidopteran and coleopteran pests have all been successfully controlled by commercial formulations of *Beauveria bassiana* and *Metarhizium anisopliae.* (Table 1.4).

Table 1.4: List of commercially available fungi-based biopesticides.

Fungi	Host range	Products
Beauveria bassiana	Effective against a variety of insect pests, including fire ants, rickets, white grubs, plant bugs, grasshoppers, flea beetles, mites, mosquito larvae, aphids, fungus gnats, whiteflies, and stored pest	Botani Gard, Naturalis, Racer BB, Ostrinil, Brocari l, Mycotrol, Boverin, Naturalis-L, Mycotrol-0 BotaniGard, Naturalis- H&G Naturalis-T&O, Mycotrol,
Metarhizium anisopliae	Effective against a wide range of insect pests. Black vine weevil, locusts and grasshoppers	Bio-Blast, Bio-Path, Green Muscle, Metaquino Pacer.
Trichoderma viride	Effective against Rot disease	Ecosom TV, Tricon, Trieco
Trichoderma harzianum	Effective against a variety of soil pathogens	Rootshield, BioTrek, Supresivit

Mode of Action of Biopesticides based on Fungi

The targeted pest dies as a result of the fungi's attack because their mycelium penetrates the insect's cuticle and enters the hemocoel. Conidia, or fungal spores, can cling to the host's cuticle due to their hydrophobic adhesion. Conidia germination happens when a germ tube grows filamentously or when an appressorium forms. The cuticle is penetrated by fungi using hydrolytic enzymes. Additionally, they release poisons, organic acids, secondary metabolites, antimicrobial substances with detoxifying effects, and secondary metabolites. The insect dies when fungus spores cross the bug's cuticle and enter its hemocoel. The internal mycelium then continues to develop and eventually breaks through the insect corpse's outer layer, where it sporulates.

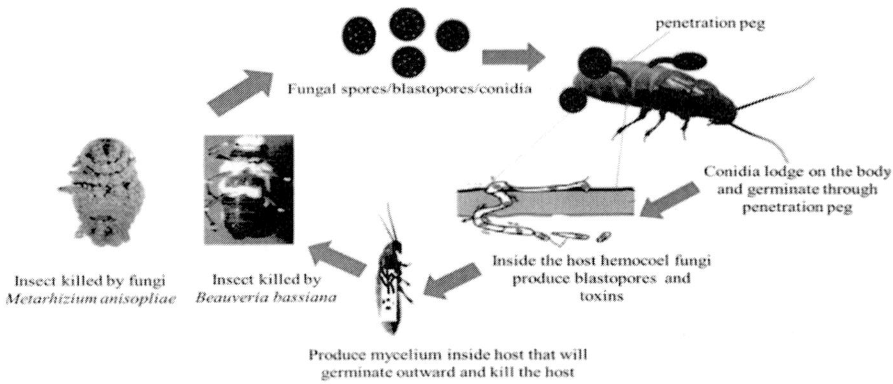

Fig. 3. Mode of action of fungi-based biopesticides

Benefits of Biopesticides Derived from Fungi

1. Wide range of hosts: Fungi are more capable of infecting a variety of pests than bacteria and viruses, including those found in fields, storage facilities, and soil.
2. Ease of mass production with low substrate requirements is possible with commercially important fungi like Beauveria, Metarhizium, Lecanicillium, and Isaria.
3. Outstanding efficacy and target specificity: Fungi-based biopesticides exhibit extraordinary efficacy and specificity.
4. Ecologically sound: The nature of these biopesticides is ecologically sound.

Disadvantages of Fungi-Based Biopesticides

1. Ideal environmental requirements: Fungi are quite sensitive to their surroundings. Arthropod cuticle penetration and spore germination both require humidity levels of 80% or greater.
2. Both temperature and UV exposure affect the survival of fungi.
3. Difficulties exist in the creation and stabilisation of fragile conidia or robust resting phases.
4. In comparison to other microbiological techniques, it is a more expensive choice.

The Utilization of Entomopathogenic Nematodes (EPN) as Bio-pesticides

Unsegmented roundworms with a soft body known as entomopathogenic nematodes are naturally present in the water layer that surrounds soil particles. These worms, which are either facultative or obligatory parasites, use a variety of sensory cues to find their hosts, including vibrations, chemical signals, and carbon dioxide. Two nematode families, Steinernematidae and Heterorhabditidae, have been effectively identified and used as biological controls in the field of pest management programmes (Kachhawa, 2017). Products including *S. carpocapsae, S. thermophilum, H. bacteriophora, and H. indica* are registered and marketed in India.

Mode of Action of EPN

1. The mouth, anus, or breathing holes of an insect serve as natural entry points for the third stage juvenile non-feeding infected insects (0.4–1.5 mm long).
2. Nematodes enter the insect's body cavity and discharge symbiotic bacteria into the intestine of the host. The toxins these bacteria create cause the insect to die within a week.
3. The nematodes liquefy the host's remnants as they feed on the dead body. On the host cadaver, they multiply and create three generations of offspring.
4. The infectious juvenile nematodes eventually leave the dead host insect after infecting it and actively look for a new host. Nematode-infected pest stages ought to appear about 5-7 days following administration under ideal circumstances. Steinernematidae cause the death of insects that are brown or tan in colour, while Heterorhabditidae cause the death of red-colored insects. However, it's not always possible to see the presence of dead insects (Fig. 1.5).

Table 1.5: List of commercially available EPN based biopesticides

Entomopathogenic nematodes	Host range	Commercial products
H. bacteriophora	Effective against root weevils, cutworms, fleas, borers and fungal gnats. Effective against black vine weevil populations by 56–100%	BioSafe, Larvanem, Nemaplant, NemaShield-HB, Nematop, Nematech-H NemaTrident-H, NemaTrident-C, Nema-green, Optinem-H
H. indica	*H. armigera, Conogethes punctiferalis, Athalia proxima*	Soldier, Nema power, Grub Terminator, Grubcure, Calterm, Aarmour
H. downesi	*Black vine weevil*	*NemaTrident-CT*

Steinernema feltiae, *S. riobravis*	Effective against black vine weevils, strawberry root weevils,	Biosafe, Ecomask, Hortscan, Guardian, Millenium, Nematac C,
	cutworms, cranberry girdler and termites	NoFlea, Savior WG, Scanmask, Termask, Vector
S. carpocapsae	Borer beetles, caterpillars, crane fly, moth larvae, coconut rhinoceros beetle	Capsanem, Carpocapsae-system, Exhibitline SC, Optinem-C, NemaGard, Nemastar,
		NemaTrident-T, NemaRed, Nemasys-C, Palma-life

Advantages of EPN

1. They are regarded as safe for people, plants, and animals as well as the environment. Handling them doesn't necessitate the use of safety masks, personal protective equipment, or re-entry intervals. They also don't leave any leftovers.
2. These biopesticides can be used to manage the cranberry girdler, root weevil, black vine weevil, webworms, cutworms, armyworms, and wood-borers.
3. Nematode survival around cranberry roots depends on maintaining wet, humid soil conditions and guarding against harmful UV rays and severe temperatures.
4. Nematodes show resistance to a variety of chemical and biological insecticides, fungicides, herbicides, fertilisers, and growth regulators when exposed for brief periods of time (between 2 and 24 hours).

Drawbacks of EPN

1. Nematodes are difficult to regulate because they are extremely sensitive to environmental conditions like UV radiation and high temperatures.
2. Nematodes are typically more expensive than standard chemical pesticides. In a natural system, however, the population of nematodes and insect hosts typically maintains a balance.
3. If the worm population rises above this natural equilibrium, the insects may be wiped out, but the nematodes will then start to drop in number until the insect populations rise again. Therefore, nematode reapplication is necessary.
4. Nematodes are unable to infect insects that are mostly found on plants because they are unable to survive above the soil. Nematode formulations also only have a short shelf life.

5. Early in the morning or late at night treatment is advised to protect the nematodes from UV light.
6. In order to hydrate the soil and make it easier for nematodes to travel into the soil, pre- and post-irrigation are both required.

Use of protozoans as biopesticides

Microsporidians, commonly referred to as protozoans, are ubiquitous intracellular parasites that can infect lepidopteran and orthopteran insects. They can therefore be used in programmes for integrated pest management. These protozoans include Nosema sp. and Vairimorpha sp. as examples. Despite the fact that they are pest-specific The use of protozoa as biopesticides has not been as successful as that of other species like bacteria, viruses, and fungi because they do not have the same pervasive and incapacitating effects on their targets. A list of successful commercially viable protozoan-based biopesticides can be found in Table 1.6.

Mode of Action of Protozoans

1. To contract microsporidia, insects must eat them.
2. The microsporidian parasite preys on the Ostrinia nubilalis, also known as the European maize borer.
3. The microsporidia spores germinate in the midgut after ingestion, and the sporoplasm is then injected into the midgut cells.
4. The spores then spread to numerous tissues and organs, proliferating and causing tissue damage as well as the emergence of the disease known as septicaemia (Senthil-Nathan, 2015).

References

Bhattacharyya A., Duraisamy P., Govindarajan M., Buhroo A.A., Prasad R. (2016). Nano-biofungicides: emerging trend in insect pest control. In: Prasad R (ed) Advances and applications through fungal nanobiotechnology. Springer International Publishing, Switzerland, pp. 307–319

Bravo A., Gill S.S., Soberon M. (2007). Mode of action of Bacillus thuringiensis Cry and Cyt toxinsand their potential for insect control. Toxicon 49(4): 423–435

Gelernter W., Schwab G.E. (1993). Transgenic bacteria, viruses, algae and other microorganisms as Bacillus thuringiensis toxin delivery systems. Bacillus thuringiensis, an environmental bio-pesticides theory and practice. John Wiley & Sons, New York, pp. 89-104

6

Biological Control of Insects

***Renuka S. Mahajan*[1]*, Devanand R. Bankar*[2]*, Pawankumar A. Sahane*[3] *and Pushpalatha M.*[2]**

[1]*Department of Entomology, Dr. Balasaheb Sawant Konkan Krishi Vidyapeeth, Dapoli-415712, Maharashtra, India*

[2]*Department of Entomology, Mahatma Phule Krishi Vidyapeeth, Rahuri-413722, Maharshtra, India*

[3]*Department of Plant Pathology, Dr. Balasaheb Sawant Konkan Krishi Vidyapeeth, Dapoli-415712, Maharashtra, India*

Introduction

The term "biological control" was first used by Smith in 1919 to signify the use of natural enemies (whether introduced or otherwise manipulated) to control insect pests. DeBach (1964) defined biological control as the action of parasites, predators, or pathogens that maintain another organism's population density at a lower average than would occur in their absence (Dhaliwal and Arora, 2016). According to a recent definition by Eilenberg *et al.* (2001) biological control or biocontrol is "The use of living organisms to suppress the population density or impact of a specific pest organism, making it less abundant or less damaging than it would otherwise be." The definition clearly states that 'living organisms' are used.

The main aim of biological management is to effective suppression of pest population by using the living agents. Biological control, in their opinion, is a process in which one species population reduces the numbers of another species through processes such as predation, parasitism, disease, or competition. Biological management is one of the best options for the sustainable management of insect. Some invasive species pest or non-native pests – species are travels from one place to another and their natural enemies are left behind, to control them and enable their spread, biological control has been an effective way to control invasive species or non-natives (CABI, 2022). Biological management should be understood as an integral component of IPM and an important part of pest-suppressing methods.

History of Biological Control

First time insect used as predators by Chinese citrus growers were red ants, *Oecophylla smaragdina*on mandarin plants to suppress leaf eating insects in 900 A.D. Italian named Aldrovandi observed that hymenopteran parasite, *Apanteles glomeratus*, laying eggs in the pupae of the cabbage butterfly, *Pieris brassicae*, in 1602. An Indian mynah bird *Gracula religiosa* was transported from India to Mauritius in 1762, to suppress the red locust, *Nomadacris septemfasciata*. The Vedalia beetle, *Rodolia cardinalis*, was introduced to control the cottony cushion scale, *Icerya purchasi*, on oranges from Australia to California (USA) in 1888.

Coccinellid lady bird beetle, *Cryptolaemus montrouzieri* was brought from Australia into India in 1898 to suppress the coffee green scale, *Coccus viridis* pest. In India (Tamil Nadu) during 1929, *Rodalia cardinalis*, a Vedalia beetle, was introduced to reduce cottony cushion scale, *Icerya purchase*. In 1937, North American species *Aphelinus mali*, was introduced to suppress the apple woolly aphid, *Eriosoma lanigerum* into Tamil Nadu. *Spogossia bezziana*, a tachnid fly, was brought into India from Sri Lanka to suppress the coconut blackheaded caterpillar, *Opisina arenosella* in 1960 (Prasad, 2019).

Why Biological management is necessary?

Biological management is an important practise in IPM. Bio pesticides are typically more target-specific and less hazardous to insects, birds, animals, and mammals than broad-spectrum conventional pesticides. They are often biodegradable in the natural environment, decreasing pollution and the likelihood of pests acquiring resistance to them. These compounds have high host specificity, usually limiting themselves to a few closely related insect species. These agents have no effect on other insect groups (natural enemies), plants, animals, or people. It is a more environmentally friendly and effective method of plant protection.

Major Characteristics of Natural Enemies

i) They should kill, reduce the rate of reproduction, inhibit growth, or shorten pest life.

ii) They are usually specific to target species or to specific damaging life stages.

iii) Their effectiveness may be affected by environmental conditions or host abundance.

iv) The degree of control by naturally occurring pathogens may be variable.

v) They are relatively slow acting; they may take several days ormore to provide sufficient management.

vi) They have the potential to generate epizootic outbreaks.

vii) Microbial are non-toxic to humans in the conventional sense, safety precautions should be followed to minimize exposure.

viii) They are non-toxic to animals and Eco- friendly in nature.

Naturally Occurring Bio-Control Agent

Natural biocontrol is the reduction of a pest population by its natural enemies, without human intervention. The insect has several naturally occurring 'natural enemies' or 'farmer's friends', these biological control agents are organisms that feed on insect's egg, larval, pupal and adult stage.A major challenge is to develop conditions that allow these beneficial organisms to reach their full potential.

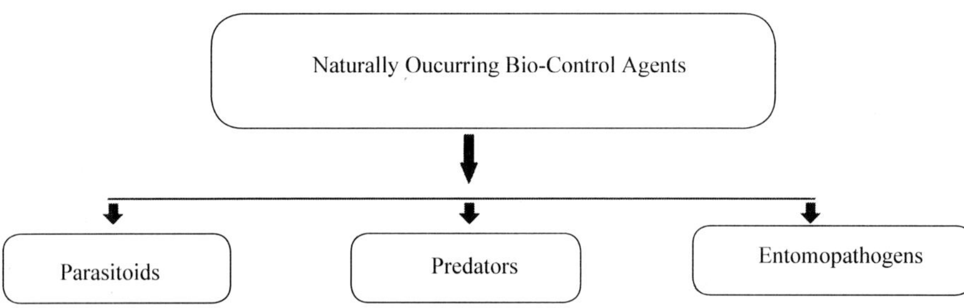

Parasitoids

Parasitoids are free-living organisms whose adults lay eggs inside or attached to a single host organism and their immature stage develops by feeding on or in a single insect host after which they are fully grown and pupate and eventually kill the host. The adults may also feed on other resources, such as honeydew, plant nectar or pollen. The parasitoids must be adapted to the life cycle, physiology and defences of their hosts species so that they have narrow host range and many are highly specialized such co-evolved parasitoids can exert a strong impact on populations of the insect and are thus good candidates and accurate identification of the host and parasitoid species is critical in the use of biological management programmes.

Types of Parasitoids (Reddy, 2018)

1. Egg parasitoids

i) Trichogrammatidae: Hymenoptera

T. chilonis used against cotton bollworm, sugarcane internode borer and rice leaf folder

T. japonicum used against rice yellow stem borer

T. acheae used against Lepidopteran pests

ii) Scleoniidae: Hymenoptera

Telenomus remus used against Tobacco caterpillar

iii) Evaniidae: Hymenoptera

Evania appendigaster targets ootheca of cockroach

iv) Eulophidae: Hymenoptera

Telastichus pyrillae used against sugarcane pyrilla

2. Egg – larval parasitoids

i) Braconidae: Hymenoptera,

Chelonus blackburni targets eggs of spotted bollworm

ii) Encyrtidae: Hymenoptera

Copidosoma koehleri targets Potato tuber moth

3. Larval parasitoids

i) Ichneumonidae: Hymenoptera

Eriborus trochanteratus used against coconut black headed caterpillar

Campoletis chloridae targets larvae of H. armigera

ii) Braconidae: Hymenoptera

Bracon hebetor and *Bracon brevicornis* used against coconut black headed caterpillar.

Cotesia plutella used against larvae of diamond back moth.

iii) Bethylidae: Hymenoptera

Goniozus nephantidis targets coconut black headed caterpillar

iv) Platygasteridae: Hymenoptera

Platygaster oryzae attacks larvae of rice gall midge

4. Larval Pupal Parasitoids

i) Ichneumonidae: Hymenoptera

Isotimajavensis attacks pre-pupal parasite of top shoot borer of sugarcane

Xanthopimpla punctata targets pupae of coconut black headed caterpillar

ii) Eulophidae : Hymenoptera

Trichospilus pupivora and *Tetrastichus Israeli* targets pupae of coconut black headed caterpillar

iii) Chalcididae : Hymenoptera

Brachymeria nephantidis attacks pupae of coconut black headed caterpillar

5. Nymphal and Adult parasitoids

i) Aphelinidae: Hymenoptera

Encarsia formosa attacking cotton whitefly

Encarsia perniciosi attacking Sanjose scale, *Quadraspidiotus perniciosus*

*Aphelinus mal i*feeds on Apple wooly aphid

ii) Epiricanidae: Lepidoptera

Epiricamia melanoleuca targets *Pyrilla perpusilla*

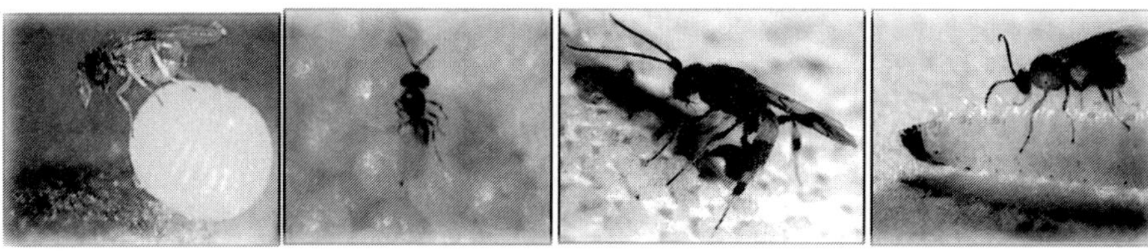

T. chilonis Telenomus remus Chelonus blackburni Bracon Sp.

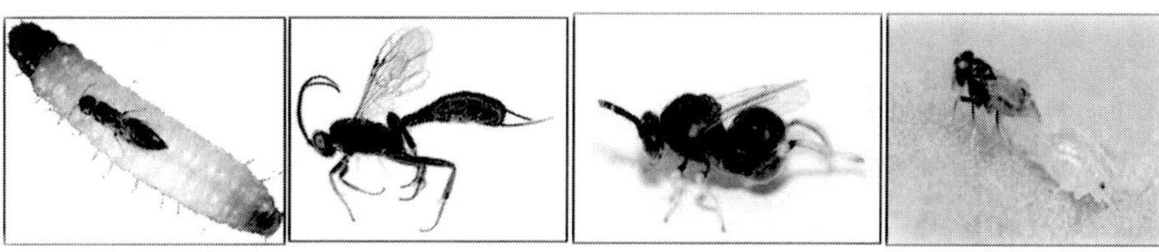

Goniozus nephantidis Campoletis chloridae Brachymeria nephantidis Aphelinus mail

Types of Parasitism (Prasad, 2019)

1. Simple parasitism

Simple parasitism refers to a single parasitoid attack on the host, regardless of the number of eggs laid. Ex: *Goniozus nephantidis* on the larvae of coconut black headed caterpillar.

2. Hyper parasitism

A parasitoid that preys on another parasitoid. It is parasitism in the form of a parasite. In actually, the hyper parasite is a secondary parasite. These reduce the primary efficiency of primary parasite. Ex:

i) The moth Hyphantria (primary host) is firstly parasitized by the dipteran-tachnid larvae, *Ernestia* (primary parasitoid) and it is, thereafter parasitized bythe chalcid larvae, *Perilampus*, which is a hyper parasitoid.

ii) Aphid is primarily parasitized by the ichneumon, Aphidius (primary parasite) and after wards parasitized by the cynip gall fly, Charips or chalcid, Asaphesfly, which are hyper parasitoids.

iii) *Aphanogmus fijiensis* is a hyper parasitoid on *Cotesia flavipes* cocoons, which isa primary parasitoid of *Chilo partellus*.

iv) The cabbage butterfly larvae *Pieris* (primary host) are parasitized first by hymenopteran *Apanteles glomeratus* (primary parasite) and it thereafter, parasitized by another hymenopteran- Natrrasticus rape which is a hyper parasitoid.

3. Multiple parasitisms

Parasitism by different species of parasitoids on the same host at a time is called multiple parasitisms. Ex:

i) *Bracon brevicornis*, *Eriborus trochanteratus* and *Goniozus nephantidis* attack larvae of black headed caterpillar at the same time.

ii) Chalcids, Eurytomamonemae and *Chrysisshanghaiensis* attack the fully grown caterpillar *Monema flavescens* at the same time.

iii) Fruit fly, *Dacus cucurbitae* larvae is parasitized first by the braconid, *Opius flatcheri* larvae and later on by the 2nd parasite chalcid. *Tetrastichus giffardianus*.

iv) Parasitization of lepidopteran eggs by two primary parasitoids, *Trichogramma* sp. and *Telenomus* sp. at the same time.

4. Super parasitism

Many individuals of the same species of parasitoid attack a single host at a time is called super parasitism. Ex:

i) *Trichospilus pupivora* on the pupae of black headed caterpillar

ii) Aphid parasitized by the braconid, Aphidius larvae

iii) Caterpillar Artona is parasitized by the larvae of Chalcid, Elachertus

5. Clepto parasitism

A type of parasitism in which the adult parasitoid preferentially appropriates for its own progeny the previously paralyzed and parasitized host of another parasitoid.

Ex: Pine shoot moth attacked by either *Eurytoma pini* or *Rhyacionia buoliana*

6. Auto parasitism

It is also called as *Adelpho* parasitism. A special type of parasitism in which the female develops as a primary parasitoid, but the male is a secondary parasitoid through females of its own species.Ex :

i) Males of Coccophagoides utilis develop as hyper parasite from unfertilized egg laid by parent female of *C. utilis* a primary parasitoid of purple citrus scale, *Parlartoria oleae*.

ii) Chalcidoid, *Encarisa formosa* females develop on the whitefly while males become internal hyper parasites of the larvae females of their own species.

7. Polyembryonic parasitism

It is the phenomenon of an individual host by single larvae of a single parasitic species, but develops into several parasitoids with in the host body

Ex: Potato tuber moth attacked by *Copidosoma koehleri*

Principles and Procedures of Biological Control (Prasad, 2019)

Principles and procedures of biological control include 4 steps.

1. Introduction of natural enemies
2. Colonization
3. Augmentation
4. Conservation

1. Introduction of natural enemies

The natural enemy should be imported from location with similar climate conditions. It should be free of other hosts. It must be free of hyperparasites and other biological competitors to the hosts. Physiological suitability of the natural enemy to its host and suitability of the host plant ecosystem. The dispersal ability of an imported natural enemy should be limited.

2. Colonization

Colonisation is a mechanism for manipulating an organism to establish it in a new location. Colonisation is the most important activity in biological control, and its success is dependent on it.

The colonisation location should be completely sheltered from pesticides and cultural practises. To reduce activity, the natural enemy should be stored at a low temperature (16°C). Most parasitoids and predators regard adult stages to be more appropriate than juvenile stages because adults can quickly adapt to new environments and are less impacted by biotic and abiotic stresses. Natural enemies should be released in the field near an appropriate host. Natural enemies should be delivered in a container with suitable relative humidity. To avoid direct heat, they must be released early in the morning or late in the evening. Before releasing natural enemies, they should be fed and mated. The number released should be determined by the natural enemy's ability to find hosts. Recovery attempts are made soon after the original release to determine if the released natural enemies are alive and under control.

Evaluation of natural enemies

Before evaluating the results, a minimum of three years is required. The efficiency of the natural enemy is assessed by comparing the pest's host population before and after release; correlating natural enemy and its host population fluctuations; contrasting the population on plots with and without a natural enemy.

They are two crucial systems that are commonly used to assess the efficiency of a natural enemy.

i) Procedures for conducting experiments (Natural enemy exclusion method)

ii) Analytical techniques (application of the life table analysis approach)

Exclusion of natural enemies can be accomplished using the exclusion method by hand picking chemical barriers of insecticides and the employment of ants. Life and mortality, as well as the causes of death (life table), are determined through analytical techniques based on numerous environmental elements for distinct stages. The data is analysed to determine the relationship between pest density and variations in density, as well as the quantity of death caused by each measured mortality component.

3. Augmentation

It involves all activities aimed at increasing the number or impact of current natural enemies. These goals can be met by introducing more natural enemies into a system or byaltering the environment to encourage higher numbers or effectiveness. These releases differ from the introduction in that they must be repeated on a regular basis. Periodic releases are possible.

I. Inoculative releases

These can be created only once a year to re-establish natural enemies. Control is expected to originate from the progeny and later generations, rather than from the release itself.

II. Inundative releases

It includes mass culturing and release of natural enemies to directly decrease the pest population. These are helpful against pests that have a single or a few distinct generations each year. The control in this case comes from the release itself.

Conservation

The most commonly employed biological control method in IPM is conservation, which is defined as activities taken to preserve and increase natural enemies through environmental management. It is possible to conserve bio-control agents via.

I. Rationalizing use of pesticides

Pesticides repel or reduce the longevity, fertility, and searching abilities of natural enemies and reduce the population of flying adults. These can be prevented by using non-persistent pesticides, skipping row treatments with pesticides, using safe and selective pesticides, using the least recommended dose, applying at specified times in a limited area, and using microbial pesticides instead of synthetic pesticides.

II. Providing food and shelter

Nesting boxes and bird perches provide suited areas for predators to gather. Plants that produce pollen and nectar are planted on the bunds to offer predators with refuge and additional food. Because natural enemies multiply in unsprayed niches, parasitoids might be introduced there to form reservoirs.

III.Beneficial management practices

- Chisel ploughing protects natural enemies of soil-dwelling bugs. The ability of entomopathogenic fungal infections to be more effective is enhanced by manipulating relative humidity and wetness periods.
- Organic manure use protects predaceous arthropods and soil-borne microbial antagonists of insect pests. If crop wastes are not burned, some parasitoids may be preserved. Parasitoids should be released when hyperparasites activity is minimum.
- Retaining crop stubble and grass weed heaps helps to maintain the predator populations.
- Keep a portion of the refuge in net mesh cages to allow parasitoids to emerge in the field.
- Overwintering parasitoids or their eggs are collected and housed in cages, and parasitoids that emerge from these cages are used early in the season.

IV. Suppression of ants

Ant suppression or physical banishment increases the efficiency of predators.

V. Determining the need for action

The natural enemy's time and period of attack should synchronize with the availability of an appropriate stage of the pest. The necessity for intervention must be evaluated for each crop ecosystem based on the population of natural enemies and their ability to feed or parasitize. When parasitoids are active predators, no parasitoids should be released; similarly, when entomo pathogenic are active parasitoids, no parasitoids should be released for effective treatment.

VI. Influence of plant types

The pubescence of plant leaves influences searching efficiency, parasitism, oviposition, and prey consumption. The colour of the leaf and flower bud attracts or repels parasitoids. By offering shelter and food, varieties with more blooms, a larger canopy, and more nectaries boost the efficacy of natural enemies. The morphological traits of the host plant, as well as the changes in volatile content profiles between cultivars, have a crucial impact in the predators' and parasitoids' efficiency.

VII. Trap cropping

Growing trap crops alongside the main crop boosts the overall efficiency of parasitoids and predators.

Predators

Predators are free-living organisms that must consume various preys to complete their life cycle. They prefer to feed on prey that is smaller than them. Insect predators include coccinellid beetles, ground beetles, anthocorid bugs, syrphid flies, and predatory mites. Coleoptera, Neuroptera, Hymenoptera, Diptera, Hemiptera, and Odonata are the most common insect orders that prey on other insects and mites. Moreover, some kinds of mites and spiders feed on a wide variety of insects and mites. Some predators, such as praying mantids, dragonflies, and beetles, utilise biting or chewing mouth parts to eat their prey, whilst others, such as assassin bugs, lacewing larvae, and hover fly larvae, use piercing and sucking mouth parts to feed on their prey's body fluids. Sucking feeders frequently inject a potent poison that swiftly immobilises their prey (Dhaliwal and Arora, 2016).

Adult predators might be monophagous (only feed on one species), oligophagous (feed on several species), or polyphagous (feed on many species). Polyphagous predators, if extremely mobile, may be efficient in pest management in disturbed environments, whereas monophagous or oligophagous predators tend to be connected with prey in more stable systems. While many predators are carnivorous as juveniles and adults, others also eat on nectar, honeydew, or plant food. Female predators frequently deposit their eggs close to their prey. Predators frequently utilise visual and chemical cues to locate plants with prey, while prey pheromones may also be used as cues. Predators frequently concentrate on more common prey species, which may make them less efficient at controlling a pest at low densities but may make them especially successful at suppressing pest outbreaks (Dhaliwal and Arora, 2016).

Coccinella septumpunctata

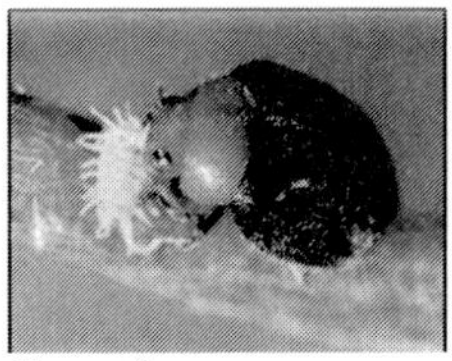
Cryptolaemus montrouzieri

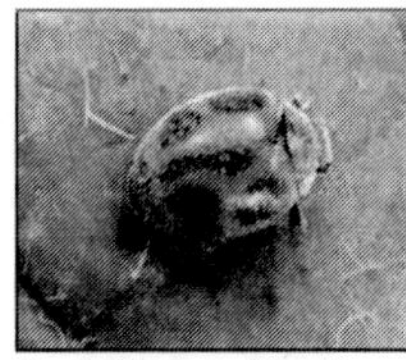
Rodolia cardinalis

Menochilus sexmaculata

Ground Beetle

Platymeris laevicollis

Green lace wings

Cyrtorhinus lividipennis

Preying mantids

Damselflies

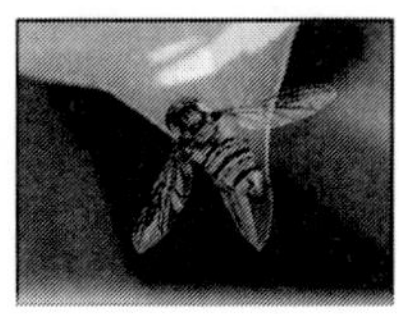
Hover flies

Robber flies

Predators used in Biological control of Insect Pests (Reddy, 2018)

Predators	**Pest Targeted**	**Predator family**
Order: Coleoptera		
Coccinella septumpunctata	Aphids	Coccinellidae (lady bird beetles)
Scymnuscoccivora	Grapevine mealy bug	
Cryptolaemus montrouzieri	Grapevine mealy bug	
Rodolia cardinalis	Cottony cushion scale	
Menochilus sexmaculata	Mealybugs and scales	
Chilocorisnigrita	*Melanispis glomerata*	
Ground beetles	Coconut BHC and rice brown plant hopper	Carabidae
Tiger beetles	Small caterpillars	Cicindellidae
Order: Hemiptera		
Platymerislaevicollis	Coconut Rhinoceros beetle	Reduviidae
Cyrtorhinuslividipennis	Rice brown plant hopper	Miridae
Eucantheconafurcelleta	Red hairy caterpillar	Pentatomidae
Order: Neuroptera		

Green lace wings or aphid lions: *Chrysoperla zastrowisillemi*	Aphids, scales and mealy bugs and boll worms	Chrysopidae
Ant lions	Soft bodied insects	Myrmeliontidae
Order: Odonata		
Damselflies (Naiads and Adults)	Caterpillars	-
Order: Mantodea		
Preying mantids (Naiads & Adults)	Caterpillar& grasshoppers	Mantidae
Order: Diptera		
Robber flies (Adult)	Small insects	Asilidae
Hover flies (Adult)	Small insects	Syrphidae

Entomopathogens

Entomopathogens are microorganisms that cause disease in arthropods such as insects, mites, and ticks and play a vital role in their control. Several naturally occurring bacteria, fungi, nematodes, and viruses infect and cause disease in a range of arthropod pests; this disease may lower the feeding rate, growth, and metabolic activity of insect pests, as well as slow or prevent reproduction, or kill them. Diseases can reproduce and spread spontaneously across an insect population under specific environmental conditions, particularly when the insect population density is high. (Stoner). Some entomopathogens are commercially produced *in vitro* (bacteria, fungi, and nematodes) or *in vivo* (nematodes and viruses).The use of entomopathogens as biopesticides in pest management is known as microbial control, and it can be an important aspect of integrated pest management (IPM) against a variety of pests (Mary Barbercheck, 2009). The active ingredient in this form of pesticide is a microorganism such as a virus, fungus, protozoan, or bacterium. Each microbe has a unique active ingredient for controlling certain pests, yet microbial pesticides can control or kill a wide range of pests that harm agricultural production. One fungus can control weeds while another can control or kill insects and pests (Nazir *et al.*, 2020).

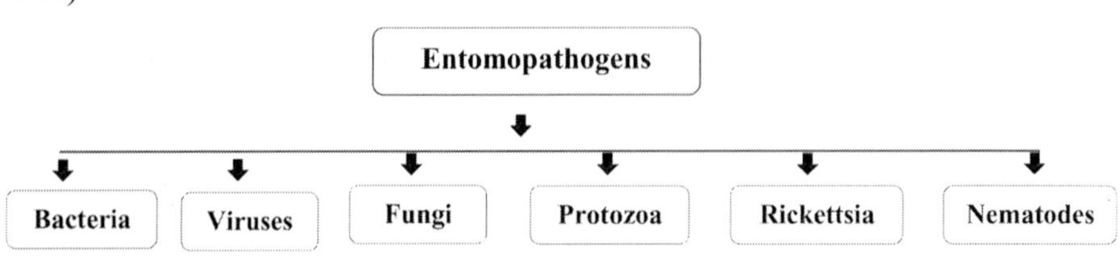

1. Bacteria

Entomopathogenic bacteria are classified as spore formers and non-spore formers (Ex: *Pseudomonas* spp.), which are further separated into obligate spore formers (Ex: *Bacillus popillae*) and facultative spore formers. The facultative spore formers could be crystal forming or non-crystalliferous (for example, *Bacillus cereus*). Spore-forming bacteria are more commonly used in pest control than non-spore-forming bacteria. Because of the deadly nature of the crystals they create, the crystalliferous spore formers are more commonly utilised for pest control than the non-crystalliferous spore formers. *Bacillus thuringiensis* (*Bt*) is the most studied crystalliferous bacteria example.

Ex. *Bacillus thuringiensis* var. *kurstaki* used against Lepidoptera

Bacillus thuringiensis var. *galleraie* used against wax moth

Bacillus thuringiensis var. *sandiego* and Bacillus thuringiensis var. tenebrionis used against beetles and weevils

2. Viruses

Baculoviruses are the most frequent and extensively researched group of viruses that are harmful to insects. Except for *Autographa californica* nuclear polyhedrosis virus, Baculoviruses are species or genus specific. Viruses are obligatory parasites, which mean they require a host to multiply. These are mostly made up of double-stranded DNA. A polyhedron-shaped protein coat protects an infective baculovirus particle (virion). The presence of polyhedral shaped viral occlusions (polyhedra) containing randomly occluded viral particles distinguishes the genus NPV. Polyhedroses are distinguished by the production of polyhedron-shaped inclusion bodies in the host's diseased tissues. Granuloses virus (GV) has one nucleocapsid per envelope and granule-shaped viral occlusions (capsules) that contain one or, in rare cases, two or more virions.Non-occluded baculoviruses have one nucleocapsid per envelope and no occlusion structures. Baculovirus infection begins when an insect consumes virus particles on a plant. *Oryctes rhinoceros* virus infection causes Malayan illness in the rhinoceros beetle, *Oryctes rhinoceros*.

Ex. *Amsacta albistriga* and *A. moorei* NPVs used to suppress red hairy caterpillars on groundnut

Ha – NPV used against *Helicoverpa armigera* on chickpea and groundnut

SI-NPV is used to control *Spodoptera litura* on tobacco, cotton and black gram

Chilo infuscutellus GV is used to against shoot borer on sugarcane

3. Fungi

Entomopathogenic fungi (EPF) reproduce both sexually and asexually in their insect hosts. In its asexual form it produces spores known as conidia which are wind dispersed. Once they are released they may land upon another insect host, or once again return to the soil where they reproduce sexually. Insect-pathogenic fungi usually need moisture to enable infection, and natural epizootics are most common during wet or humid conditions. The effectiveness of these fungi against pest insects depends on having the correct fungal species and strain with the susceptible insect life stage, at the appropriate humidity, soil texture (to reach ground-dwelling pest species), and temperature. The fungal spores, which can be carried by wind or water, must contact the pest insect to cause infection. Unlike bacterial and viral pathogens of insects, fungal pathogens infect the insect with contact and do not need to be consumed by their host to cause infection.

Ex. *Beauveria bassiana* causes a disease to European corn borer, codling moth, cabbage caterpillar and whitefly and the disease known as the white muscardine.

Metarhizium spp. causes a disease known as green muscardine. It is effective against rootworm, white grubs, and some root weevils.

Verticillium lecanii is used against whitefly and thrips and aphids.

Nomuraea rileyi is used against green clover worm, cabbage looper, cabbage worm, armyworms, corn earworm, and tobacco budworm.

Hirsutella thompsonii and *Neozygites floridana* are used against mites.

4. Rickettsiae

The pathogenic rickettsiae differ mainly from bacteria in that they are Gram negative, obligate intracellular pathogens with typical bacterial cell walls and no flagella. Rickettsia is a genus of non-motile, Gram-negative, non-spore forming, highly pleomorphic bacteria that can present as cocci (0.1 um in diameter), rods (1- 4 μm long) or thread - like (10 μm long). Obligate intracellular parasites, the Rickettsia survival depends on entry, growth, and replication within the cytoplasm of eukaryotic host cells (typically endothelial cells). Because of this, Rickettsia cannot live in artificial nutrient environments and are grown either in tissue or embryo cultures.

Ex. *Rickettsiella popilliae* is used against Japanese beetle causing blue disease. *Rickettsiella melolontha* is used against scarabaeid beetles.

5. Nematode

Entomopathogenic nematodes (EPNs) are soil-inhabiting, lethal insect parasitoids, live inside the body of their host, and so they are designated endoparasitic. The most commonly studied genera are those that are useful in the biological control of insect pests, the Steinernematidae and Heterorhabditidae.

Ex. *Steinernema capocapsae* used against Lepidopterous larvae.

Heterorhabditis bacteriophora used against Soil-dwelling insect

6. Protozoa

The pathogen enters the insect body through the gut wall, spreads to different tissues and organs, and multiplies, sometimes causing tissue breakdown and septicemia. These are relatively slow acting organisms, taking days or weeks to debilitate their host. Protozoa kill the insects either directly or by reducing the fecundity of the adult. Their effect on host is chronic. They prolong the larval life in the field, thus exposing the insect longer to predators and parasitoids. These are called debilitating infections. They are always associated with other pathogens.

Ex. *Vairimorpha necatrix* is used against corn earworm and European cornborer, various armyworms, fall webworm, and cabbage looper.

Nosema melolonthae is used against chaffer beetles.

Farinocystis triboli is used against red flour beetle.

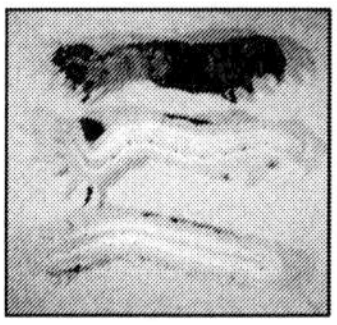

Bt infected larvae

Fungi infected larva (Green muscardine)

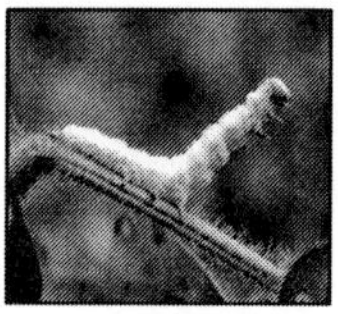

Fungi infected larva (White muscardine)

NPV infected larva

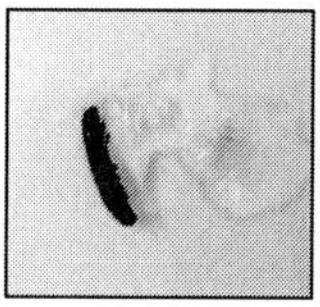

Nematode infected larva

Conclusion

Biological control of insect pest means the use of natural enemies such as parasitoids, predators and entomopathogens to suppress their population and damage caused by them. The most important and commonly used insect parasitoids include *Trichogramma* sp., *Telenomus* sp., *Chelonus* sp., *Bracon* sp., *Goniozus* sp., and *Epiricamia melanoleuca.* Ladybird beetles, ground beetles, tiger beetles, green lacewings, Preying mantids, reduviid bug, damselflies, robber fly, and hoverflies are key insect predators. Spiders and some mite groups are also insect and mite pest predators. Entomopathogens are disease-causing bacteria, fungi, protozoans, and viruses. Natural enemies were introduced into the environment and their potential was expanded by encouraging them to multiply. Natural enemies are more likely to survive when a non-toxic control strategy is utilised. Bio-control is an important part of integrated pest management and helps to reduce the use of synthetic chemical pesticides. Biocontrol of insect reduces risk of resistance development in insect, high efficiency, low cost, eco-friendly, safety for beneficial organisms, reduction of residues in environment, target-specific in nature and increased biodiversity in human managed ecosystems.

References

CABI (2022). Biological control (bioprotection) beginners guide.

DeBach, P. (1964). Biological Control of Insect Pests and Weeds. Cambridge University Press, Cambridge, UK, pp. 844

Dhaliwal, G.S. and Arora, R. (2016). Integrated Pest Management concepts and approaches. Kalyani Publishers, New Delhi, pp. 201-234

Eilenberg, J., Hajek, A. and Lomer, C. (2001). Suggestions for unifying the terminology in biological control. BioControl, 46(4): 387–400. doi: 10.1023/a: 1014193329979

Mary E. Barbercheck (2009). Biological Control of Insect Pests. eOrganic article

Nazir, T., Khan, S. and Qiu, D. (2020). Biological Control of Insect Pest.Pest Control and Acarology, pp. 1-14. DOI: 10.5772/intechopen.81431

Prasad, T.V. (2019). Handbook of Entomology. New Vishal Publications, New Delhi, India, pp. 178-234

Reddy, D.S. (2018). Applied Entomology. New Vishal Publications, New Delhi, India, pp. 150-155

7

Role of Entomopathogenic Microorganism in Pest Management

Shailja Jaiswal[1], *Sudheer Kumar*[2], *Archana Upadhyay*[1]
Girijesh Kumar Jaiswal[1] *and Ajit Pandey*[3]

[1]*Department of Entomology, Chandra Shekhar Azad University of Agriculture and Technology, Kanpur-208002, Uttar Pradesh, India*

[2]*Department of Entomology, P.G. College, Nawabganj, Gonda, Uttar Pradesh, India*

[3]*Department of Entomology, Banda University of Agriculture & Technology Banda-210001, Uttar Pradesh, India*

Introduction

The paper summarises recent findings on nematodes, viruses, bacteria, and other entomopathogenic microorganisms (EM) and their potential application in biological pest control. It's important to carefully research the ecological characteristics of EM and their function in nature in order to utilise various biological control tactics against insect hosts efficiently. Since it identifies frequent patterns in the interactions between insect pests and EM that have just lately been uncovered through research, this chapter aids readers in understanding the significant advancements in this sector. Weeds, arthropods (mainly insects and mites), mollusks (slugs and snails), some vertebrates, and plant pathogens (oomycetes, fungus, bacteria, nematodes, and viruses) are only a few examples of the diverse species that are considered agricultural pests. These pests reduce agricultural output and product quality by munching on crops. It is estimated that there are millions of pest species in the globe, and they have a significant detrimental influence on agricultural output, decreasing ideal world crop yields by 40%. Insect pests and mites serve as hosts for a number of naturally occurring fungus, bacteria, nematodes, and viruses that infect a range of organisms. Insect pests alone are thought to be responsible for 10.80% of all agricultural losses experienced globally in the years after the "green revolution." Up to 80% of the time, the intensive use of pesticides has resulted in the development of insecticide resistance to one or more modes of action. Entomopathogens, which include bacteria, viruses, protozoa, and fungi, are therefore thought of as controlling factors in pest infestations.

Insect population management relies heavily on naturally occurring entomopathogens. Tanzini *et al.* (2001) used the term "entomopathogens" to refer to microorganisms that control insect pest populations to levels that do not result in any financial harm to crops. Delgado and Murcia (2011) expanded the definition of the term to include populations of microorganisms that prey on pests by invading them, disrupting their life cycles, and using them as hosts. These microbes were categorised as either facultative or obligate parasites with high survival potential that prey on insect pests. In addition to being successful against insect pests, using entomopathogenic microorganisms (EM) as pest management agents is also environmentally safe, with reduced pesticide residues for both people and non-target creatures. Natural enemies are crucial in preventing pests in an agro-ecosystem from reaching the economic injury level (EIL). Depending on the type of pest population and the biological features of the control agent, biological control agents (BCAs) are applied in a variety of methods. BCAs have a number of enticing qualities, including host specificity, the ability to control toxic residues, the lack of phytotoxic effects, safety for humans, and the possibility for long-term insect pest control. However, the natural enemy and the ecology of the pest are necessary for effective BCA use. It is necessary to create effective biological control agents using entomopathogens (viruses, fungus, bacteria, and nematodes), which calls for a complete understanding of bioassay techniques as well as manufacturing, formulation, and application methods.

Entomopathogenic Fungi

Worldwide, entomopathogenic fungi (EPF) are essential to biological pest management. EPF are heterotrophic, eukaryotic microorganisms that may reproduce sexually and asexually, exist as unicellular or multicellular (filaments), and produce a variety of infectious propagules. Environmental factors including UV radiation, temperature, and humidity have an impact on the effectiveness of EPF in the field. Among those with the greatest EPF (Entomophthoromycota) diversity are the orders Onygenales (Ascosphaera genus), Entomophthorales, Hypocreales, and Neozygitales. Important entomopathogenic taxa include Beauveria, Metarhizium, Nomuraea, Entomophthora, Verticillium, and Neozygites, which are found in a variety of taxonomic groups. EPF can have an impact on insects from the orders Coleoptera, Lepidoptera, Hemiptera, Orthoptera, Diptera, and Hymenoptera. Few fungi, like those in the family Hypocreales, can potentially live on a variety of hosts, but others, like those in the genus Entomophthorales, are only found on certain kinds of insects. Lepidopterous larvae, aphids, and thrips are examples of entomopathogenic fungi that have been seen to infect several mite and insect pest species. These pests are significant economic agricultural pests in many countries. Due to their host-specificity, EPF rarely harm non-target species while developing deadly illnesses and controlling insect and mite populations in nature.

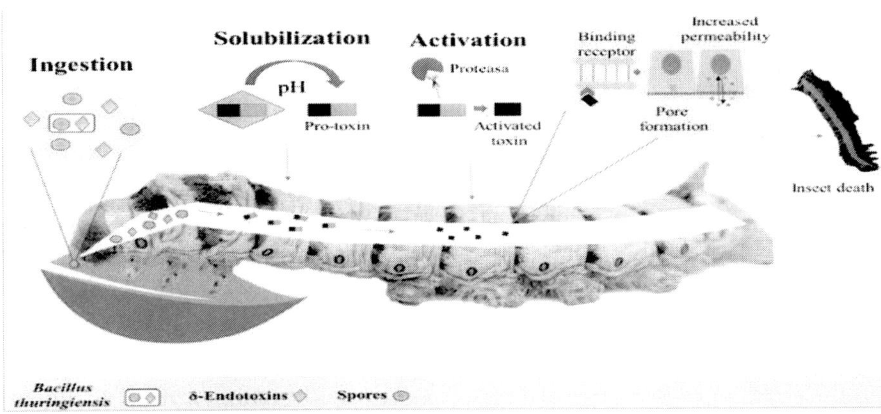

The appressoria, which are specialised structures, are generally covered in a mucus layer of proteins and glucans and are created when the spores adhere to the insect cuticle. The mechanical pressure and enzyme activity (lipases, proteases, and chitinases) in the germ tube are what cause the insect cuticle to penetrate. Most EPF develop vegetatively inside the hemocoel of the insect. Toxins created and released by an infection or mechanical harm caused by mycelia developing inside an insect (mummification) are the main reasons of insect mortality. Toxins such destruxin, bavericin, and efrapeptins are secreted by *Beauveria*, *Metarhizium*, and *Tolypocladium*; their actions and contributions to the pathogenesis process are well characterised. The fungus continues its life cycle on new hosts after the insect dies by producing hundreds to thousands of fresh spores on the deceased carcass. Bemisia tabaci discovered that the host plant had an impact on the pathogenesis of the fungus *B. bassiana* (Hypocreales: Cordycipitaceae) to *B. tabaci. B. bassiana* is known for its pathogenicity to the sweet potato. The mortality brought on by *B. bassiana* as well as the development of newly generated conidia were significantly influenced by the host plant on which the nymphs were raised.

Entomopathogenic Bacteria

In general, biological control techniques, such as using bacterial entomopathogens, are thought to be safer than chemical pesticides and promise enormous financial gain. Their mode of action is often more complex than that of conventional pesticides, and they tend to target a variety of locations where resistant bugs are more likely to develop. Entomopathogenic bacteria can be used as a stand-alone pest management tool, but to achieve optimal efficiency and ecological safety, they work best when used in rotation or in conjunction with insecticides. Numerous studies have shown that synthetic pesticides and enteropathogenic microorganisms work well together. Incorporating biopesticides into pest management programmes has the added advantages of protecting workers, reducing crop residues, allowing for more flexibility in harvesting due to a short or nonexistent pre-harvest time, and protecting the environment. Families include Bacillaceae, Streptococcaceae, Pseudomonadaceae, Enterobacteriaceae, and Micrococcaceae are home to the bulk of bacterial diseases that affect insects. Some of these bacteria are extremely virulent, but the majority of them are soft pathogens that spread to stressed insects and infect or afflict them. Bacillaceae has been the subject of the most speculation among them. While *Bacillus sphaericus* is effective against mosquitoes, Bacillus *popillae* causes milky sickness in scarabaeids. In order to control caterpillars and beetles, *Bacillus thuringiensis* (Bt.) is frequently employed as an entomopathogenic agent. Bt. is a bacterium that forms spores, and the generation of proteinaceous protoxin crystals with insecticidal characteristics is correlated with the spores' creation. In the insect's stomach, the eaten crystals disintegrate and release endotoxin, an active pollutant. When consumed, special insecticidal endotoxins known as Cry proteins generate pores that harm the insect stomach epithelium. Bt. creates parasporal (crystal-like) structures with these endotoxins. Cry toxins that are marketed commercially have shown to be effective against a variety of insect pests, including as Lepidoptera, Coleoptera, and Diptera species, with little to no direct impact on unintended creatures. A study on the use of genetically modified Bt to suppress *Leptinotarsa decemlineata* (Coleoptera: Chrysomelidae) revealed no effect on non-target organisms such predatory Hemiptera. Bt. bioinsecticides applied to agricultural ecosystems and other habitats typically do not boost spore production, and their viability tends to decline, especially when exposed to sunshine. A microbial control treatment called Bt is sprayed on insect infestations to quickly kill the target insects and, in most cases, prevent them from reproducing. It's crucial to take the risk to species other than the targets into account when assessing the security and ecological effects of EPBs. Some plants have the Bt-endotoxin-encoding genes, making them resistant to a variety of insect pests. According to farmer surveys, growing Bt crops greatly reduces the need for conventional pesticides. However, some problems have emerged, such as worries about the mirid insect on Bt cotton in China. A few years after Bt cotton became widely used and failed to control the mirid pests that were previously controlled by a wide range of insecticides, problems with the pests began to arise in China. Since they became commercially available, genetically modified (GM) maize and cotton plants that express lepidopteran-active endotoxins have transformed agriculture in the nations where they are grown. Currently, GM crops are grown in eight countries: Brazil, Paraguay, Argentina, South Africa, the United States, Canada, and China. GM crops do have certain drawbacks.

Entomopathogenic Viruses

Entomopathogenic viruses (EPV), commonly referred to as insect-killing viruses, have recently come to be recognised as important global contributors to biological pest management. The first virus-based insecticide to control the cotton bollworm was approved in the USA in 1970, despite the fact that several viruses were investigated for controlling insect pests in the early 1900s. There are now several viruses that are authorised for use in the management of insect pests, and continuing research attempts to find and assess additional viruses (López-Ferber, 2020). Numerous insect orders have been shown to contain EPVs, which can target a wide variety of insects. They serve as effective biological control agents, especially when used against insect pests that might contract viral diseases. EPVs can include either RNA (dsRNA and ssRNA) or double- or single-stranded DNA (dsDNA and ssDNA). Since the sixteenth century, insect pathogens known as EPVs have been reported. They have proven effective against a variety of plant pests in agro-ecosystems around the world. Numerous viral illnesses, including nucleopolyhedrosis, have been identified in honeybees (Apidae, Hymenoptera: Bombycidae), Bombyx mori L., silkworms, and other Lepidopteran species. Viral particles are made up of nucleic acids encased in a protein covering called a capsid, which is crucial to the host cell's infection process. Once a virus has subtly entered a cell, its nucleic acid seizes control of the host metabolic system and begins to produce numerous copies of itself until the cell inevitably perishes. Because they are obligatory parasites, viruses cannot grow in vitro. The 12 viral families into which EPVs are divided by the International Committee on Taxonomy of Viruses (ICTV). Three of them—Polydnaviridae, Baculoviridae, and Ascoviridae are insect-specific families that are extremely host-specific and nonpathogenic to beneficial insects and other non-target animals, including mammals. Particularly the Baculoviridae family has been regarded as an environmentally safe substitute for chemical insecticides. There are two types of baculovirus (dsDNA) in the family Baculoviridae: granulovirus (GV) and nucleopolyhedrovirus (NPV), which are particularly important for suppressing lepidopteran pests on crops. The genomes of both groups are circular double-stranded DNAs. In both insect cells and human cells, the baculovirus has been extensively employed to make a variety of recombinant proteins. In terms of external symptoms, each group of EPVs exhibits different manifestations. The insect's activity slows down, it stops eating, and its growth stops as the first obvious signs. Every EPV family has a different method of pathogenesis and reproduction, but infection is virtually always spread through ingestion. The virus passes past epithelial cells and combines with receptors in the insect's stomach. After reaching the hemocoel, infection spreads to critical tissues and organs, especially fat bodies. Within 5 to 14 days, severe infections cause the host to pass away. Insects infected with baculoviruses have a white appearance as a result of a serious infection of the fat body, which causes the integument (exoskeleton) to gradually shrink until it ruptures, releasing billions of occlusion bodies (OBs) that help inocula multiply in the field. Four nuclear polyhedrosis virus isolates from the beetroot armyworm *Spodoptera exigua* H. (Lepidoptera: Noctuidae) have been the subject of studies. The isolates (SeNPVSP1 and SeNPVSP2) were from the USA, Thailand, and Spain. The modest restriction fragment length polymorphism in the viral genomes indicated related but different genotypes (variants). As a constraint fragment length polymorphism marker, the Bg III fragment of each isolate was employed. The genome of SeNPVs is 134 kbp in size. The mobility patterns of the occluded virion polypeptide and polyhedrin are very similar between the four SeNPV isolates. After being incorporated by *Staphylococcus aureus* V8, the polyhedrin from SeNPVUS was determined to be distinctive. SeNPV showed the lowest LD50 in the second instar *S. exigua* larvae bioassays, with just 1.5 polyhedra per second instar larva. With cotton pests including *S. exigua* and *P. gossypiella* (Lepidoptera: Gelechiidae), as well as the Heliothis or Helicoverpa complex, EPVs have demonstrated efficacy. Commercial baculovirus biopesticides are thought to present a low risk to both people and wildlife because they typically have a limited host range and are typically confined to the order and family of the host of origin. Only in vivo can baculoviruses be created in large quantities, but they are commercially viable for more advanced hosts like Lepidoptera.

Entomopathogenic Nematodes

Entomopathogenic nematodes (EPNs) are roundworms with a soft body and no segments that are obligatory parasites of insects. They are about 0.5 mm in length and come from 23 different worm families. Heterorhabditidae and Steinernematidae, two EPN families, have been used successfully in pest control programmes as biological pesticides. These nematodes are found naturally in soil and hunt for their insect victims in response to chemical cues like CO2. EPNs made for sale are utilised as biological control agents for a variety of pest insects that live in soil. EPNs and bacteria interact in a mutualistic symbiotic way that aids in the biological control of insect pests. Since they are easily mass-produced and deployed using standard spray equipment, they are environmentally benign and adapt well to a variety of situations. Infectious juveniles enter the hemocoel of their host insect once inside and release symbiotic bacteria that have been stored in the nematode's gut. These bacteria cause septicemia, which kills the host in 24 to 48 hours.

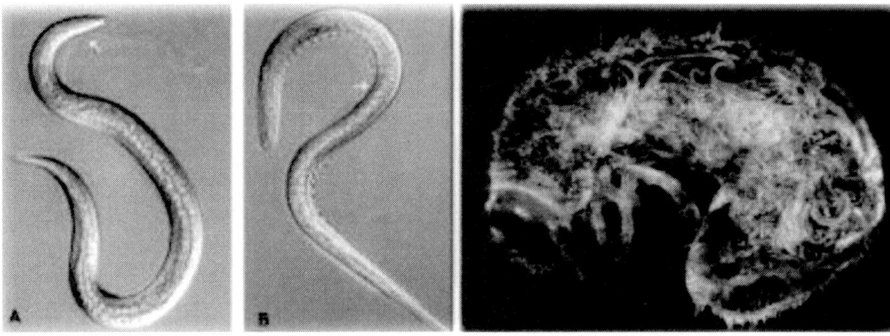

Then, infectious juveniles feed on the bacterial populations that are expanding quickly and the rotting host tissues. The nematode goes through two to three generations inside the host corpse. As bacteria serve as food, the symbiotic connection with bacteria promotes nematode proliferation and pathogenicity. By introducing bacteria into the host, nematodes serve as vectors that provide the necessary circumstances for nematode survival and reproduction in the insect body. EPNs are well known for having the potential to be used as a biological control agent in plant protection, particularly those that belong to the Steinernematidae and Heterorhabditidae families. All Heterorhabditis nematode species are connected to Photorhabdus bacteria, but all *Steinernema* spp. are connected to *Xenorhabdus* bacteria. Heterorhabditids and steinernematids are widely distributed in soils all over the world. Their usefulness against different pest insects has been well studied. The mass manufacturing of EPNs can be carried out via in vivo or in vitro techniques. For example, *Galleria mellonella* larvae (Lepidoptera: Pyralidae) are frequently utilised as a substitute host to develop nematodes, and a liquid fermentation technique is used to produce nematodes on a big scale. In order to cultivate EPNs in big quantities for use in industry, huge fermenters are employed to expose worms to a pure culture of their symbiont in a nutritive media. Nematode virulence and viability tests, age, and the proportion of viable to non-viable worms are used to assess the nematode product's quality. EPNs enter the hemocoel of their host insects by the spiracles, mouth, anus, or, in the case of some species, the intersegment membranes of the cuticle. They inject symbiotic bacterial spore into the sick insect's hemolymph, where the bacteria quickly grow and induce septicemia, which kills the host within 24 to 48 hours. Nematodes develop, mature, and breed while still feeding on the dead host tissue. After 8–14 days, a new generation of infectious juveniles appears. The only free-living stage for EPNs is the infectious juvenile stage. Heterorhabditis and Steinernema are mutualistic partners with the bacteria Photorhabdus and Xenorhabdus, respectively. The juveniles expel symbiotic bacterial cells into the hemocoel from their intestines. Infected insects die as a result of the bacteria reproducing in their hemolymph. The nematodes continue to grow, reproduce, and eat host tissue. A distinct reproductive mechanism is used by heterorhabditid nematodes, which generate both men and females in the following generation after infectious juveniles develop into hermaphrodite adults. In contrast, all generations of steinernematid worms produce both men and females. EPNs do best in sandy soil with a pH range of 4–8, although they are sensitive to UV radiation, cold, dehydration, and excessive heat.

Plant pathogens such as fungus, viruses, oomycetes, bacteria, and nematodes, weeds, arthropods (mostly insects and mites), molluscs (slugs and snails), and a few vertebrates are examples of agricultural pests. By eating crops, these pests reduce crop quality and yield. Pest species are thought to number in the millions worldwide, severely reducing agricultural productivity. In fact, it has been reported that these failures have resulted in a 40% reduction in the potential world crop output. Insects and mites are among the agricultural pests that are particularly difficult to control and account for roughly 10.80% of agricultural losses globally in the post-green revolution age. Due to their convenience, high efficacy, and ease of use, insecticides are now a vital tool for controlling insect pest infestations and have been widely employed to reduce these losses. However, the intensive use of chemicals has resulted in the development of insect pest resistance, with as much as 80% of cases observing resistance to one or more classes of insecticides. Entomopathogens, which include fungi, viruses, protozoa, and bacteria, are thought of as natural pest infestation regulators in order to deal with this problem. Entomopathogens manage insect pest populations and keep them at levels that don't hurt crop plants financially. The word "entomopathogens" refers to microorganisms that prey on insect pests as a necessary component of their life cycles, defining them as obligatory or facultative parasites with a high possibility for survival. In addition to being effective against insect pests, using entomopathogenic microorganisms (EM) as pest control agents is also environmentally safe for both people and non-target creatures because it decreases pesticide residues. In agro-ecosystems, natural enemies are essential for keeping pest populations from rising to dangerous levels. Depending on the type of pest and the biological characteristics of the control agent, biological control agents (BCAs) are used in a variety of ways. Their host specificity, lack of toxic residues, lack of phytotoxic effects, safety for humans, and potential for self-sustaining pest control are only a few of their benefits. However, effective application necessitates a thorough comprehension of both the ecology of the insect and its natural antagonist. A full grasp of bioassay processes, manufacturing, formulation, and application strategies is required for the development of entomopathogens (fungi, bacteria, viruses, and nematodes) as effective biological control agents.

Conclusions

Chemical pesticides are frequently used to protect plants, however this strategy has increased insect resistance to several chemical compounds found in these products. Exploring all-natural approaches to combating insect infestations, such as entomopathogens, has received more attention in recent years. Microorganisms known as entomopathogens hunt out and kill insect pests, providing hope for efficient management of insect pests. Entomopathogens are being developed as environmentally benign substitutes for agricultural crops because they can act as biological control agents to improve agro-sustainability. It is commonly acknowledged that using biological controls to protect agricultural crops from insect pests is a responsible ecological strategy. The study of microbial pesticides offers a singular opportunity for performing prospective and predicative research within the field of pesticides.

References

Karabörklü, S., Azizoglu, U. and Azizoglu, Z.B. (2018). Recombinant entomo-pathogenic agents: a review of biotechnological approaches to pest insect control. World Journal of Microbiology and Biotechnology, 34: 1-12

Ruiu, L., Satta, A. and Floris, I. (2013). Emerging entomopathogenic bacteria for insect pest management. Bull Insectol, 66(2): 181-186

Skinner, M., Parker, B.L. and Kim, J.S. (2014). Role of entomopathogenic fungi in integrated pest management. Integrated Pest Management, 169-191

Sujeetha, J.A.R. and Sahayaraj, K. (2014). Role of entomopathogenic fungus in pest management. Basic and Applied Aspects of Biopesticides, 31-46

8

Conservation Techniques of Predators in Agro-ecosystem

Garima Tak[1] *and Anshul Sharma*[2]

[1]*Department of Entomology, Shrinathji College of Agriculture, Nathdwara, Rajasthan*

[2]*Department of Agronomy, Rajasthan College of Agriculture, Udaipur, Rajasthan*

Introduction

Providing sufficient habitat and other food supplies to already-existing or naturally-occurring beneficial species is one of the most successful and long-lasting ways for reducing economic harm from pests. In contrast to monocultures or pesticide-treated fields, beneficial species such as predators, parasites, and pest-sicking "pathogens" are significantly more numerous in varied farms where pesticides are used less frequently.

What are Predators?

Predators: Insects are consumed by a diverse range of predators. Many creatures rely extensively on insects for food, including birds, amphibians, reptiles, fish, and mammals. These insectivorous animals eat a wide variety of insect species and rarely focus on pests unless they are in great numbers. Insects and other arthropods are more commonly used in biological control because they consume a restricted range of prey species and have shorter life cycles than other arthropod predators. This is due to the fact that their population density can shift in response to variations in the density of their prey. A few notable insect predators are lady beetles, ground beetles, rove beetles, flower bugs and other predatory true bugs, lacewings, and hover flies. Spiders, among other things, are insect predators and pests. A predator makes its living by catching and eating on other animals. Predators are often bigger and stronger than their victim. Many of the most prevalent predators in fruit production systems attack a wide variety of pest species and aid in pest population density regulation. Some examples of predators in the agro-ecosystem are-

Ladybugs, also known as lady beetles or ladybird beetles, are a beloved pest control for farmers and gardeners. They are known for their ability to gobble up crop-destroying aphids, which are a significant issue for farms and gardens.

Ladybugs can consume 50-60 aphids daily and thousands over a lifetime, and can lay hundreds of eggs in a season. They also feed on mites, scales, mealy bugs, thrips, and white flies, which all damage crops.

Spiders are extremely valuable to humans because they suppress insects and crop-eating pests. They are major insect controllers, averting starvation and suppressing crop-consuming insects. Spiders are unsettling because of their sticky webs and fast pounces, yet they are necessary for a healthy environment.

Coyotes, a pest animal and a nuisance, may assist farmers by being skilled predators that control rodent populations. They are omnivorous and consume fruits, berries, and vegetables, although meat accounts for 90% of their diet. They may also attack and kill animals in packs, which ranchers may find offensive. Good fence and a livestock guard dog or two roaming the zone are usual solutions.

Skunks are a Grade A nuisance that may help farms by eating problem insects such as caterpillars, grubs, field mice and voles. They consume berries, leaves, and grasses as well. Skunks, on the other hand, are not attracted to beekeepers due to their thick fur since they are not deterred by bee stings and would seek out honeybee colonies for food.

Owls are super-carnivores that help farmers reduce rodent infestations and safeguard crops. They are expert nocturnal hunters, taking up to 10% of their total weight on rodents each night. Their enormous wingspan enables them to pursue phantom night stalkers, and they have also been observed preying on snakes, amphibians, and invertebrates. Owls' ability to track and hunt at night makes them a natural pest defence system.

Lacewings attack aphids, spider mites, whiteflies thrips, leafhoppers, scales, mealybugs, psyllids, small caterpillars and insect eggs. Green lacewing larvae feed on insect pests. Both larvae and adult brown lacewings feed on pests. Green lacewings are commercially available.

Syrphid (Hover) fly larvae attack Aphids, scales, thrips and other small soft-bodied insects

Praying mantids attacks many, including aphids, flies, beetles. Feeds on pests as well as beneficials.

Mantids grasp their prey with spined front legs and hold them while they eat.

Parasitic wasps attacks aphids, caterpillars and whiteflies.

Predatory mites attacks spider mites, thrips, fungus gnat larvae.

Role of predators in agro ecosystem

Pests are those creatures that prey on a resource that humans desire to conserve and are successful enough in doing so to cause either economic harm or just general aggravation. They make up a very small portion of the local bug species. Even many of the species that we would consider to be severe pests only seldom harm us or our resources in a substantial way.

Predators are crucial in reducing the populations of potential pests. When insecticides have wiped off the predators of prospective bugs, this has been consistently shown. When insects are liberated from the care of their predators, they frequently turn into destructive pests, even though they were previously of little economic value. On the other hand, when a non-toxic technology is discovered to treat a significant pest, the reduced use of pesticides and increased survival of predators frequently reduces the numbers and damage of formerly important secondary pest species.

Avoid farming practises that kill beneficial predators in order to conserve and enhance rich predator populations. Instead, use measures that will help them survive. Begin by discontinuing practises that interfere with natural biological management, such as pesticide sprays, hedge clearance, and extensive herbicide usage to remove weeds in and around fields. During critical times of the growing season, even little changes in farming practises can drastically increase natural enemy populations. Simple applications of straw mulch provide damp, safe hiding places for nocturnal predators such as spiders and ground beetles. Straw mulch may also reduce the capacity of flying pests such as aphids and leafhoppers to "see" the crops they attack by decreasing the contrast between foliage and bare soil. Insect damage to mulched vegetable plots can be significant.

Diversification has several advantages, as is true of the majority of the tactics discussed in this book. For instance, correctly chosen blooming plants or trees on field borders may serve as significant sources of beneficial insects, but they may also alter the microclimate of crops, contribute to organic matter, and yield timber or feed. Creating wild flower borders around agriculture fields increases the number of helpful insects that are actively seeking for pollen and nectar. The beneficials then relocate to nearby fields to assist in controlling insect problems. Many of these flowers also have the added benefit of being good bee food, which increases honey output, or they may be sold as cut flowers, which increases farm income.

Methods to conserve predators

Increase the number of predators: A fertilised, weeded, and watered monoculture is a dense, pure concentration of an insect pest's favourite diet. Over time, many people have adapted to these simple cropping techniques. Predators, on the other hand, do poorly since they are suited to natural systems. Tilling, weeding, spraying, harvesting, and other common farming practises degrade desirable habitat. Instead, try to meet their biological demands.

Predators require more than prey and hosts to complete their life cycles; they also require refuge places and alternate meals. Many adult parasites, for example, feed on pollen and nectar from adjacent flowering weeds while looking for hosts. Predaceous ground beetles, like many other predators, do not disperse far from their overwintering locations; access to fixed food sources is essential.

Provide more resources: Increase predator populations by giving supplies to attract or retain them on your farm.

Allowing permanent populations of alternate prey to vary below harmful levels increases the survival and reproduction of helpful insects. To do this, put plants that host alternate prey surrounding your fields or even as strips within your fields. The relative quantity of aphids in cabbage influences the efficacy of general predators that devour diamond back moth larvae. Similarly, anthocorid bugs benefit from alternate meals when their primary prey, western flower thrips, is limited in many areas.

Another technique is to increase the number of the favoured host of a beneficial bug. Cabbage butterflies, for example, are the preferred host for two parasites (Trichogramma evanescens and Apanteles rebecula). With continuous releases of pregnant females, populations of this pest increased roughly tenfold in the spring. This allowed the two parasites — both parasitic wasps — to establish themselves sooner in the spring and remain active throughout the season. Because of the apparent hazards, this approach should only be used in instances where pollen, nectar, or alternate prey cannot be acquired.

Manage the vegetation in field margins: With proper preparation, you may transform your field margins into predator reservoirs. These areas can be key overwintering grounds for crop pest predators. They can also give pollen, nectar, and other nutrients to predators.

Manage plants near fields to control certain pests: One method, known as perimeter trap cropping, works best when plants such as snap beans or cowpeas are cultivated to keep stink bugs and Mexican bean beetles away from soybeans. Plants that are highly appealing to target pests are planted around a cash crop, completely surrounding it without gaps, in perimeter trap farming.

Pesticide treatments may be drastically reduced by enticing pests away from the cash crop via perimeter trap cropping. By restricting pesticide application in field edges or eliminating it entirely, we can preserve the predators.

Create predator corridors: By seeding varied blooming plants into strips every 165 to 330 feet (50-100 m) throughout the field, you may supply predators with highways of habitat. Beneficiaries can circulate and disperse into field centres by using these corridors.

Pick the proper plants: Beneficial insects are drawn to certain plants, so if you're attempting to manage a certain pest, pick blooming plants that will attract the suitable beneficial insect(s). The size and form of the blooms determine which insects may reach the pollen and nectar of the flowers. The most beneficial blooms, including parasitic wasps, are little and somewhat open. Plants of the aster, carrot, and buckwheat families are particularly beneficial.

Use weeds to attract beneficials: While weeds are sometimes the best blooming plant for attracting beneficials, this practise complicates maintenance. Although some weeds promote insect pests, harbour plant diseases, or compete with cash crops, others provide critical resources to many beneficial insects and contribute to agroecosystem biodiversity.

Unquestionably, weeds may stress crops, yet there is strong evidence that farmers can increase beneficial insect populations by modifying weed species and weed control practises. Many farmers are emphasising weed management over weed control as they get a better understanding of the intricate interactions between crops, weeds, pests, and predators.

Improve plant pest defences: A healthy plant is the first line of defence against insect pests. Plants that are healthy are better equipped to endure the onslaught of insect pests and can respond by mobilising inbred defence systems. Improve soil and provide the greatest possible growth circumstances, including appropriate (but not excessive) water and nutrients, to boost natural defences.

Plants have developed a variety of insect defences, including both continuous and triggered responses. Around the clock, continuous defences are maintained, including poisonous compounds that prevent feeding and barrier construction. Pest assaults cause induced responses, allowing plants to utilise resources more flexibly. Plant defence mechanisms that are both effective and long-lasting combine continuous and triggered responses. In reaction to insect or mite attacks, crops may either directly or indirectly emit poisonous compounds, relying on other parties for aid. Volatile chemicals attract predators, and plants under assault produce different molecules than plants that are not under attack. Different types of the same plant or sections of the same plant might produce varying amounts and proportions of volatile chemicals. Leaves that are not injured generate and emit volatiles, which increases the signalling capacity of damaged plants. As an example, when a beet armyworm chews on a cotton plant, the plant releases a specific chemical signal blend into the air, which attracts female parasitic wasps. However, inappropriate nitrogen levels can alter the chemical signal's ratio, rendering it unnoticeable.

Plant breeding: while overwhelmingly advantageous in the near term, can have unintended repercussions that derail plant geneticists' best-laid intentions.

Because the emphasis of pest resistance plant breeding is frequently restricted to a single plant/pest interaction, choosing for one resistant gene may mistakenly remove other genes influencing other pests or genes that have a role in attracting the insect's predators.

Furthermore, newly produced varieties may stand better or yield more at the price of natural defences, which are sometimes sacrificed unwittingly for those other traits. Selecting for a single feature, such as height, may result in mistakenly selecting against any of the many inborn plant defences against pests.

Preserving present predators or predators: Using less pesticides is vital since most predators are extremely sensitive to them, reducing their field efficacy greatly. The original idea behind integrated pest management (IPM) was to combine biological and chemical control by employing the fewest pesticides required for economic production and delivering those that were required in a method that did not interfere with biological control agents. Pesticide-resistant cultivars, cultural practises that reduce pest abundance or damage, approaches for modifying insect mating or host-finding behaviour, and, in certain cases, physical measures of control can all assist to reduce pesticide use. However, many IPM initiatives have failed to get beyond the early development stage.

The influence of insecticides on predator populations is determined by their physiological effects and administration strategies. Herbicides and fungicides can be harmful to beneficial insects, but synthetic pyrethroids are even more dangerous. The toxicity of systemic and ingested pesticides on predators is lower. Predatory phytoseiid mites are prevented from ovipositing by fungicides such as benomyl, thiophanate-methyl, and carbendazim. To reduce pesticide influence on predators, carefully arrange treatments, restrict contact, and avoid susceptible life stages. Spot treatments or alternating strips over a field may also be beneficial.

Supplying predators with habitat and resources: Unless they are released again every year, predators must have a suitable habitat to overwinter in as they are often dormant throughout the winter months. Others may spend the winter on agricultural wastes, other plants, or soil; however certain predators and diseases overwinter in the bodies of their hosts (who may then have their own overwintering requirements). Predacious mites overwintering in apple orchards are a well-known example. In these orchards, the ground cover offers protection from the pesticides used on the fruit trees, winter shelter, a supply of pollen, and alternative prey.

During the summer, many predators' adults may need or profit from pollen, nectar, or honeydew (made by aphids). Many agricultural plants only blossom consistently for a brief period of time, therefore additional supplies of pollen and nectar may be required from flowering plants inside the field or along its borders. However, plant diversity within a field can potentially hinder the effectiveness of host-finding. The availability of pollen and alternative food may stabilize the populations of generalist predators, but the efficiency of the predators still depends on their capacity to react swiftly, either by aggregation or multiplication, to outbreaks of the target pest. Thus, diversification of plants or other methods of supplementing the nutrition of predators must be done with knowledge of the behavior and biology of the natural enemy and pest.

For instance, the Colorado potato beetle's eggs and early instar larvae provide a viable food source for the local lady beetle *Coleomegilla maculata*. The quantity of aphid prey in nearby fields, such as alfalfa, brassicas, cucurbits, and corn crops, as well as the accessibility of pollen from corn and numerous weeds, including dandelion and yellow rocket, determines the number of potato beetles that feed on them. Despite the fact that this predator does not now effectively manage *C. maculata* numbers in the agricultural environment, it could be more successful in the future.

Creating a stable population and adding new predators: The biology of the pest, its predators, and any potential unintended consequences (such as adverse impacts on native species that are not pests or on other predators of the pest) must all be thoroughly researched throughout this phase. It is necessary to quarantine suitable predators after they have been identified, researched, and collected in order to get rid of any parasites or diseases that may have affected the population. The predators are then gradually released, paying close attention to the right moment in the enemy and pest life cycles, in an area with a high concentration of the target pest and with little disruption to the freshly released enemies. Although this process is long and complex, when it is successful, the results can be impressive and permanent, as long as care is taken in production practices to minimize negative effects on the natural enemy.

Periodic releases of predators and mass culture

Seasonal immunization releases

In certain circumstances, the weather or a lack of adequate hosts or prey prevent a natural enemy from successfully overwintering. In other circumstances, such as in greenhouses, the natural enemy's potential habitat is completely eliminated at the conclusion of the growing season or production cycle. Therefore, it may be necessary to periodically reintroduce the natural enemy to maintain pest control, especially in annual crops or other highly disturbed systems.

Biological insecticides or inundative release

Because they do not seek to create a population of predators that grows to a size where it achieves a long-term balance with the population of their hosts or prey, these two methods of biological control are fundamentally distinct from all the others. Instead, the plan is to release biological agents in sufficient numbers to reduce the insect population, much like a pesticide. The majority of insect pathogen formulations that are commercially available are frequently utilised.

Biological pesticides derived from Bacillus thuringiensis function by paralysing the gut of insects. The active element in these products is a bacterial protein, and many do not include living bacterial spores, preventing disease transmission. Beneficial nematodes are live predators that are deployed in large numbers to attack their hosts and release symbiotic microorganisms. They feed on bacteria and insect tissue, mate, and reproduce, producing new juvenile nematodes after one to two weeks. Because nematodes are very resistant to desiccation, UV radiation, and harsh temperatures, they are effective against insects in soil or protected habitats. Adequate moisture and temperatures ranging from 53 to 86° Fahrenheit are required for success. Despite the high costs of bulk raising, storage, and shipping, artificial diet research continues.

Conclusion

Predators include birds, amphibians, reptiles, fish, and mammals, among others, that devour insects. Because of their shorter life cycles and limited prey species distribution, they are more typically utilised in biological control. Predators including lady beetles, ground beetles, rove beetles, flower bugs, lacewings, and hover flies survive by catching and consuming other creatures. They are frequently bigger and stronger than their prey and play an important role in pest management in fruit production systems.

Farmers and gardeners rely on ladybirds, spiders, and coyotes to manage pests. Ladybirds eat crop-damaging aphids, mites, scales, mealy bugs, thrips, and white flies. Spiders manage bug and crop-eating pests, but coyotes are excellent predators that keep rodent populations in check. Skunks, who are classified as Grade A nuisances, eat berries, leaves, and grasses but are not drawn to beekeepers because to their thick fur. Owls, who are super-carnivores, assist farmers in reducing rodent infestations and protecting crops. Lacewings, syrphid fly larvae, praying mantids, parasitic wasps, and predatory mites are among the other insects. These insects perform an important role in keeping the environment healthy and pests at bay.

Natural predators can benefit from environmentally friendly agricultural practises by providing habitat, food, and shelter. Predators require a variety of resources, but simple efforts can assist in providing habitat for a variety of species. Invertebrates that prey on humans seek refuge in field edges, buffer strips, and neighbouring semi-natural environments. Hedgerows and rough vegetation provide cover and buffer temperature fluctuations, while flower-rich field edges provide habitat for insect predators such as the marmalade hoverfly and parasitic wasps. Through targeted pesticide application, an Integrated Pest Management (IPM) technique can generate habitat for predatory invertebrates while also reducing chemical exposure.

References

El-Wakeil, N., Saleh, M., Gaafar, N. and Elbehery, H. (2017). Conservation Biological Control Practices. InTech. doi: 10.5772/66312

El-Wakeil, Nabil. (2017). Conservation Biological Control Practices

Grandperrin, Thomas (2019). Using conservation control to aid your pests' enemies. UAV-IQ Inc Precision Agriculture

Landis, D.A., Wratten, S.D. and Gurr, G.M. (2000, January). Habitat Management to Conserve Natural Enemies of Arthropod Pests in Agriculture. Annual Review of Entomology, 45(1): 175–201. https://doi.org/10.1146/annurev.ento. 45.1.175

Larry Gut, Annemiek Schilder, Rufus Isaacs and Patricia McManus (2002). Chapter 2: Managing the Community of Pests and Beneficials. Fruit Crop Ecology and Management

Miguel A. Altieri, Clara I. Nicholls and Marlene A. Fritz (2005). Manage Insects on Your Farm, SARE Handbook Series 7

Snyder, William (2019). Give predators a complement: Conserving natural enemy biodiversity to improve biocontrol. Biological Control. 135. 10.1016/j.biocontrol.2019.04.017

9

Role of Predatory Insects in Biological Control of Crop Weed

***Abhishek Singh Chauhan*[1], *Kamesh Singh Chauhan*[2], *Sagar Bhagwan Dongarjal*[3], *Shekhar Vilas Mane*[4] and *Mahesh Kumar*[5]**

[1] *Phytosanitary Inspector, Accredited from DPPQ&S at Nasik*

[2] *M.Sc. Agronomy, CCSU Meerut, Uttar Pradesh*

[3] *Research Associate Entomology Telluris bio-tech Pvt. Ltd. Hyderabad*

[4] *Technical Officer, P.Q. Station Nasik*

[5] *M.Sc. Extension, Chandra Shekhar Azad University of Agriculture & Technology Kanpur-208002, Uttar Pradesh*

Introduction

In the relentless battle against crop weeds, farmers have traditionally relied on chemical herbicides and labour-intensive mechanical methods. However, concerns about environmental impact and herbicide resistance have driven the search for more sustainable and eco-friendly alternatives. One such solution is biological control, harnessing the power of predatory insects to manage weed populations naturally. This chapter delves into the fascinating world of biological control, exploring the key concepts, benefits, and implementation strategies of using predatory insects to combat crop weeds.

1. Understanding Biological Control

What is Biological Control

Biological control, also known as bio-control, is a method of pest management that involves using living organisms to control pest populations. Predatory insects are natural enemies of certain weed species, preying on them to regulate their numbers and limit their growth.

The Importance of Balance in Ecosystems

By embracing biological control, we embrace the principles of ecological balance. In natural ecosystems, various species coexist, with predators keeping prey populations in check. Implementing this balance in agriculture can reduce the need for chemical interventions and promote sustainable practices.

2. Predatory Insects as Weed Warriors

Identifying Beneficial Predators

Numerous predatory insects have shown an appetite for specific weed species. Ladybugs (Coccinellidae) target aphids and other soft-bodied pests, while lacewings (Chrysopidae) prey on aphids, mites, and small caterpillars. Ground beetles (Carabidae) are voracious feeders, targeting a wide range of weed seeds and small pests.

Predator-Prey Dynamics

Understanding the intricate relationship between predators and their weed prey is essential for effective biological control. Predators have developed specific adaptations to locate, consume, and control weed populations, often leading to a delicate ecological dance.

3. Implementing Biological Control

Site-Specific Approach

Each farm's unique characteristics and weed challenges call for a tailored biological control strategy. By studying the ecosystem, climate, and weed species prevalent in a region, farmers can identify the most suitable predators for release.

Mass Rearing and Augmentation

To boost predator populations and increase their efficacy, mass rearing and augmentation techniques are employed. Specialized facilities rear predatory insects in controlled conditions, ensuring they are healthy and ready for release.

4. Predator Release and Monitoring

Strategic Timing and Distribution

Releasing predators at the right time is critical for a successful bio-control program. Timely releases coincide with peak weed growth stages, maximizing the impact of the predatory insects.

Monitoring the Efficacy

Continuous monitoring is essential to evaluate the success of the bio-control program. Researchers and farmers observe weed density, predator populations, and crop health, making adjustments as needed to optimize results.

5. Integration with Other Weed Management Practices

Enhancing Effectiveness through Integration

Biological control is most effective when integrated with other weed management practices. Combining techniques like crop rotation, cover cropping, and mulching can create a comprehensive and sustainable weed control approach.

Challenges and Limitations

While biological control offers numerous benefits, it is not a one-size-fits-all solution. Challenges such as varying weather conditions, the potential for unintended effects on non-target organisms, and the need for careful planning and coordination exist.

History of Biological Control

Biological control, also known as bio-control or biological pest control, is a method of managing pest populations using natural enemies or beneficial organisms. The concept of using predatory insects for weed control dates back thousands of years, with ancient civilizations employing various organisms to manage pests and unwanted plant growth. However, the modern scientific understanding and development of biological control began to take shape in the 19th century.

Early Attempts and Notable Successes

Ancient Practices: In ancient China, farmers used ants to control citrus pests, and the Chinese also practiced the release of predatory ants to manage caterpillar pests in citrus orchards. Similarly, the ancient Egyptians employed cats to control rodent populations in granaries and storage facilities.

Medieval Europe: During the medieval period, falconers in Europe used birds of prey to control rodent populations, helping protect crops from damage.

Introduction of Ladybugs: In the late 19th century, the vedalia beetle (*Rodolia cardinalis*) was successfully introduced from Australia to California to control the cottony cushion scale (*Icerya purchasi*). This marked one of the earliest and most notable successes in the biological control of agricultural pests.

Cactoblastis Moth Control of Prickly Pear Cacti: In the early 20th century, the cactoblastis moth (*Cactoblastis cactorum*) was introduced to Australia from South America to control invasive prickly pear cacti, which were causing significant agricultural and ecological damage. The cactoblastis moth proved highly effective in controlling the cacti population.

Key Milestones and Developments in the Use of Predatory Insects for Weed Control

The Green Revolution: In the mid-20th century, agricultural practices underwent significant changes with the advent of the Green Revolution. During this period, the focus was primarily on increasing crop yields using chemical fertilizers and pesticides, which led to a decline in the emphasis on biological control methods.

The Silent Spring Effect: Rachel Carson's groundbreaking book, "Silent Spring," published in 1962, highlighted the detrimental environmental impacts of synthetic chemical pesticides. This led to a renewed interest in the development of eco-friendly alternatives, including biological control methods.

Advancements in Research and Technology: Over the decades, research and technology in the field of biological control have advanced significantly. Scientists have identified and studied numerous predatory insects with potential applications in weed control, providing a more comprehensive understanding of their behavior and effectiveness.

Integrated Pest Management (IPM): The concept of Integrated Pest Management emerged as a holistic approach to pest and weed management, aiming to combine various control methods, including biological control, for sustainable agriculture. IPM emphasizes the use of predatory insects and other natural enemies as part of an environmentally friendly and economically viable pest management strategy.

Conservation Biological Control: This approach involves enhancing the habitat and resources for natural enemies, including predatory insects, to increase their populations and effectiveness in controlling pests and weeds. Conservation biological control has gained prominence as a key component of sustainable agriculture.

Predatory Insects as Weed Control Agents

Predatory insects play a crucial role in natural ecosystems by preying on various pests, including insects, mites, and small animals. When it comes to weed control in agricultural settings, these beneficial insects can also be harnessed as effective agents to suppress weed populations. Predatory insects offer several advantages over chemical methods and can be an integral part of sustainable weed management strategies. Here's a closer look at their characteristics and benefits as weed control agents:

1. Target-Specific Feeding

Predatory insects exhibit target-specific feeding behavior, meaning they primarily consume certain pest species. This trait is advantageous in weed control as they tend to focus on particular weed species, reducing the risk of non-target damage to crops or desirable vegetation.

2. Natural Enemy of Weeds

Predatory insects have evolved to be natural enemies of certain weed species. Their biological adaptations, such as specialized mouthparts or digestive enzymes, allow them to efficiently consume and control weeds in a way that is often specific to the target weed.

Selective and Environmentally Friendly

Unlike broad-spectrum chemical herbicides, predatory insects are selective in their feeding and do not harm non-target organisms. This makes them environmentally friendly and suitable for integrated pest management (IPM) programs, which aim to balance pest control with conservation efforts.

Self-Sustaining Populations

Once established, populations of predatory insects can often self-sustain if sufficient prey or alternative food sources are available. This reduces the need for repeated introductions and can lead to long-term weed suppression.

Reduced Chemical Use

Using predatory insects for weed control can help reduce the reliance on chemical herbicides, contributing to a decrease in chemical residues in the environment and minimizing potential negative impacts on beneficial organisms.

Compatibility with Organic Farming

For organic farmers, the use of predatory insects aligns well with organic farming principles, as it is a natural and non-synthetic approach to pest and weed management.

Low Risk of Resistance

Unlike chemical herbicides, which can lead to the development of herbicide-resistant weed populations, predatory insects are less likely to encounter resistance issues due to the complexity of their feeding behaviors and the multiple mechanisms they employ to control pests and weeds.

Limited Environmental Persistence

Chemical herbicides can persist in the environment for extended periods, potentially leading to unintended side effects. In contrast, predatory insects typically have a short lifespan and may not have long-lasting effects once their prey populations decline.

Potential for Use in Various Environments

Predatory insects can be used in a range of agricultural environments, including field crops, orchards, and greenhouse settings, providing versatile options for weed control across different agricultural systems.

While predatory insects offer significant benefits for weed control, successful implementation requires careful consideration of factors such as predator selection, population management, and compatibility with existing weed management strategies. Additionally, understanding the ecology and behavior of both the predatory insects and target weed species is critical to optimizing their effectiveness in suppressing weed populations and promoting sustainable agricultural practices.

Selection and Identification of Predatory Insects

Selection and identification of predatory insects for weed control is a crucial step in implementing a successful biological control program. It involves carefully evaluating potential predators to ensure they are effective against the target weed species while being compatible with the agricultural ecosystem. Here are the key considerations and steps involved in the selection and identification process:

Identify the Target Weed Species

Begin by identifying the specific weed species that need to be controlled. Understanding the biology, life cycle, and habitat preferences of the target weed will help in selecting appropriate predators that can efficiently attack and suppress the weed population.

Research Potential Predatory Insects

Conduct thorough research to identify potential predatory insects known to feed on the target weed or related species. Consult scientific literature, research papers, and reputable resources that document the feeding preferences and effectiveness of various predatory insects against specific weed species.

Determine Predator Specificity

Assess the feeding habits and specificity of each potential predatory insect. Ideally, the predator should primarily target the desired weed species without causing harm to non-target plants, crops, or beneficial organisms.

Evaluate Environmental Suitability

Consider the environmental conditions and climate of the agricultural area where the biological control will be implemented. Ensure that the selected predatory insects can thrive and effectively control weeds under the prevailing conditions.

Assess Life Cycle and Behavior

Understanding the life cycle and behavior of the predatory insects is essential. Consider factors such as their reproductive rates, feeding patterns, and dispersal abilities. Predators with overlapping generations and a preference for the target weed throughout their life stages are more likely to provide sustained weed control.

Determine Availability and Source of Predators

Identify sources from which the predatory insects can be obtained. This could include commercial suppliers, biological control organizations, or research institutions that rear and distribute beneficial organisms for pest management.

Conduct Field Trials and Observations

Before implementing large-scale biological control, conduct small-scale field trials to assess the effectiveness of the selected predatory insects against the target weed. Monitor predator populations and weed suppression over time to gauge the success of the approach.

Consider Complementary Predators

In some cases, a combination of different predatory insect species may be more effective in controlling the target weed. Evaluate the potential for synergistic effects by using complementary predators that have different feeding behaviors or life cycle timings.

Monitor for Unintended Consequences

During and after introducing predatory insects, closely monitor their impact on the weed population and other components of the ecosystem. Look for any unintended consequences, such as non-target damage or shifts in pest populations, and adjust the management approach if necessary.

Develop an Integrated Weed Management Plan

Integrate the use of predatory insects into a comprehensive weed management plan that may include other methods such as cultural practices, mechanical weed removal, and herbicide use. An integrated approach is often the most effective and sustainable way to manage weeds populations in agriculture.

Ultimately, successful selection and identification of predatory insects for weed control require a well-informed decision-making process based on scientific knowledge, field observations, and a thorough understanding of the target weed and the predators under consideration. Collaboration with entomologists, weed scientists, and experienced biological control practitioners can significantly enhance the chances of success in implementing a bio-control program.

Augmentation and Conservation of Predatory Insects

Augmentation and conservation of predatory insects are essential components of successful biological control programs for weed management. Augmentation involves the deliberate release of additional predatory insects into the target area to boost their populations, while conservation focuses on creating and enhancing habitats to support the long-term survival and effectiveness of these beneficial organisms. Here are the key considerations for both augmentation and conservation:

Augmentation of Predatory Insects

Assessment of Pest Pressure: Before augmentation, assess the level of weed infestation and pest pressure in the area. Determine whether the existing population of predatory insects is sufficient to provide effective weed control or if additional releases are necessary.

Source and Quality of Predators: Obtain predatory insects from reputable suppliers or specialized rearing facilities to ensure that the released individuals are healthy, free from diseases or parasites, and have a high likelihood of survival and establishment.

Timing of Releases: Time the releases of predatory insects to coincide with the life stages of the target weed species. Releasing predators when the weed is most susceptible to herbivory maximizes their impact.

Distribution Strategy: Distribute the released predatory insects evenly across the infested area to ensure uniform weed control. Pay attention to areas where weed growth is most concentrated or vulnerable

Monitoring and Evaluation: Continuously monitor the population dynamics of the released predators and the impact on the target weed. Regular evaluations help to assess the success of the augmentation strategy and make adjustments as needed.

Conservation of Predatory Insects

Habitat Management: Design and manage the agricultural landscape to provide suitable habitats for predatory insects. Incorporate features such as hedgerows, wildflower strips, and cover crops to offer shelter, alternative food sources, and overwintering sites.

Minimize Pesticide Use: Reduce or eliminate the use of broad-spectrum pesticides in the target area to avoid unintentional harm to predatory insects. Employ selective and targeted pesticides only when necessary.

Diverse Plantings: Encourage the establishment of diverse plant communities within and around crop fields. A variety of plant species can attract and sustain a diverse range of predatory insects, increasing the resilience of the overall ecosystem.

Natural Enemy Reservoirs: Identify and protect areas where native predatory insects thrive naturally, such as nearby woodlands or natural habitats. These areas can act as reservoirs, providing a source of predators that can colonize the crop fields when needed.

Reduced Soil Disturbance: Minimize unnecessary soil disturbance and tillage practices, as these activities can disrupt predator populations and their overwintering sites.

Biological Corridors: Create corridors or connections between natural areas to facilitate the movement of predatory insects between different parts of the landscape, enhancing their ability to find and control weed populations.

Crop Rotation and Diversification: Implement crop rotation and diversification strategies to disrupt weed growth cycles and reduce the build-up of weed populations. This can positively impact the success of predatory insects in managing weeds.

Monitoring and Assessment

Monitoring and assessing the biological control of crop weeds through predatory insects is an essential aspect of integrated pest management (IPM) strategies. Biological control involves using natural enemies, such as predatory insects, to reduce weed populations and minimize the use of chemical pesticides. Here are some key steps in monitoring and assessing the effectiveness of biological control using predatory insects:

Identification of Target Weed and Predatory Insects

Identify the weed species that need to be controlled and the predatory insects that can potentially prey on those weeds. It's crucial to have a good understanding of the specific weed and predator species involved.

Establishment of Predatory Insects

If predatory insects are not already present in the area, consider introducing them into the crop ecosystem. This process is known as "augmentation," and it involves releasing an adequate number of predatory insects to control the weed infestation effectively.

Monitoring Weed and Predator Populations

Regularly monitor the population levels of the target weed and the introduced predatory insects. This can be done through visual observations, trapping, and sampling techniques. Monitoring should start before the release of predators to establish a baseline, and it should continue throughout the growing season.

Assessing Weed Damage

Determine the level of weed damage in the crop fields and compare it with areas where no biological control measures have been implemented. This comparison will help evaluate the effectiveness of the predatory insects in reducing weed populations.

Monitoring Predator Establishment and Persistence

Check whether the released predatory insects have successfully established in the ecosystem and if they persist over time. Factors like climate, habitat suitability, and the presence of alternative prey can influence their survival and effectiveness.

Measuring Weed Suppression

Quantify the reduction in weed populations achieved through the actions of predatory insects. This can be done using various metrics, such as weed density, weed biomass, or weed cover.

Assessing Crop Yield and Quality

Evaluate the impact of weed suppression on crop yield and quality. Healthy crops with minimal weed interference should result in higher yields and better-quality produce.

Economic Evaluation

Conduct an economic analysis to determine the cost-effectiveness of using biological control through predatory insects compared to other weed control methods, such as chemical herbicides.

Adjusting Strategies

Based on monitoring and assessment results, make necessary adjustments to the biological control program. This might include altering the release rates of predatory insects, introducing additional natural enemies, or implementing supplementary control measures if needed.

Data Recording and Reporting

Keep detailed records of monitoring and assessment results. This information will be valuable for future decision-making and for sharing the outcomes of the biological control program with relevant stakeholders.

Future Prospects and Challenges

The biological control of crop weeds through predatory insects holds significant promise as a sustainable and environmentally friendly approach to managing agricultural pests. However, there are also some challenges that need to be addressed to realize its full potential. Here are the future prospects and challenges associated with this method:

Future Prospects

Sustainable Pest Management: Biological control offers a long-term, sustainable solution for weed management, reducing the reliance on chemical pesticides and their potential negative impacts on the environment and human health

Biodiversity Conservation: The use of predatory insects for weed control can contribute to the conservation of natural biodiversity by promoting the presence of beneficial insect species in agricultural landscapes.

Resistance Management: Unlike chemical pesticides, which pests can develop resistance to over time, biological control agents can adapt to changing conditions, making them a valuable tool for managing resistant weed populations.

Integration with Other Methods: Biological control can be integrated with other pest management practices, such as cultural and mechanical control methods, to create a comprehensive and effective weed management strategy.

Cost-Effectiveness: Once established, natural predators can reproduce and persist, reducing the need for repeated applications, potentially making it a cost-effective approach in the long run.

Public Perception: Biological control is generally viewed more favorably by the public, as it aligns with the principles of sustainable agriculture and reduced chemical use.

Challenges

Host Specificity: Some predatory insects may not be highly specific to the target weed species and could also prey on non-target plants or beneficial insects. Ensuring that introduced predators focus primarily on the target weed is critical to avoid unintended consequences.

Availability of Effective Predators: Finding and sourcing effective predatory insects can be a challenge in some regions, particularly for specific weed species and certain agricultural ecosystems.

Environmental Factors: The success of biological control can be influenced by environmental factors like temperature, humidity, and habitat availability. Adverse conditions may limit the effectiveness of predatory insects.

Regulatory Hurdles: Introducing non-native predators or other biological control agents may require strict regulatory approvals, which can be time-consuming and costly.

Timing and Coordination: The successful implementation of biological control often requires precise timing and coordination with the weed's growth stage and the release of predatory insects.

Resistance to Predators: Some weed species can evolve defenses against predatory insects, reducing the efficacy of biological control over time.

Public Awareness and Education: Farmers and other stakeholders may require education and training to adopt biological control methods effectively.

Risk of Displacement: If not managed carefully, introduced predators might displace native species, affecting local ecosystems.

Conclusion

Biological control of crop weeds through predatory insects presents a promising and eco-friendly solution for modern agriculture. By embracing the natural balance of predator-prey relationships, farmers can minimize the use of chemical herbicides, protect beneficial organisms, and preserve the environment. The integration of this innovative approach with other sustainable practices empowers farmers to foster healthier crops, ensure food security, and promote the long-term well-being of our planet. As we look to the future, embracing biological control may hold the key to a more harmonious and sustainable agricultural landscape.

References

Al-Ani, L.K.T., Aguilar-Marcelino, L., Fiorotti, J., Sharma, V., Sarker, M.S., Furtado, E.L. and Herrera-Estrella, A. (2020). Biological control agents and their importance for the plant health. In Microbial services in restoration ecology (pp. 13-36). Elsevier

Cividanes, F.J. (2021). Carabid beetles (Coleoptera: Carabidae) and biological control of agricultural pests in Latin America. Annals of the Entomological Society of America, 114(2): 175-191

Cloyd, R.A. (2020). How effective is conservation biological control in regulating insect pest populations in organic crop production systems? Insects, 11(11): 744

Hinz, H.L., Winston, R.L. and Schwarzländer, M. (2019). How safe is weed biological control? A global review of direct nontarget attack. The Quarterly Review of Biology, 94(1): 1-27

Huffaker, C.B. (Ed.). (2012). Theory and practice of biological control. Elsevier

Mehdizadeh, M., & Mushtaq, W. (2020). Biological control of weeds by allelopathic compounds from different plants: a bioherbicide approach. In Natural remedies for pest, disease and weed control (pp. 107-117). Academic Press

Nazir, T., Khan, S. and Qiu, D. (2019). Biological control of insect pest. Pests Control and Acarology, 21

Turner, C.E. (2020). Tephritidae in the biological control of weeds. In Fruit fly pests (pp. 157-164). CRC Press

Van Driesche, R. and Hoddle, M. (2009). Control of pests and weeds by natural enemies: an introduction to biological control. John Wiley & Sons

10

Mass Production Techniques of Entomopathogenic Fungi and Their Biological Application for Insect Pest Management

Pankaj Singh[1], *Ravi Kumar*[1], *Deepak Mourya*[2] *and Deepak Kumar*[1]

[1]*Department of Plant Pathology, Sardar Vallabhbhai Patel University of Agriculture & Technology, Meerut, Uttar Pradesh*

[2]*Department of Plant Pathology, Shri Khushal Das University, Hanumangarh Rajasthan*

Introduction

Insect pests cause a loss of 18-26% of global yearly agricultural yields. A wide range of the losses 13-16% occur in the field prior to harvest. Furthermore, post-harvest pests are responsible for a significant portion of agricultural product storage losses. Poor management practises might result in the loss of 50-60% of stored grains over the storage period of time.Farmers have used chemical pesticides to control insect pests for many years but the effects on non-target organisms, groundwater contamination, residues on food crops and the emergence of insect pests resistant to chemical pesticides have forced scientists to concentrate on the development of alternative eco-friendly measures.Biological pest management is one of the most effective alternatives to chemical pesticides. Biological control, defined as pest population reduction by natural enemies is an alternative to manufactured chemical pesticides that generally requires human intervention (Hoffmann and Frodsham 1993).

Entomopathogenic fungi are an important aspect of an integrated strategy that can provide considerable and selective insect control among the numerous bio-control agents. Entomopathogenic fungi are fungi that kill insects by attacking and infecting their insect hosts (Singkaravanit *et al.*, 2010). They are probably the most adaptable biological control agents due to their broad host range. These fungi are a diverse group of over 100 genera with over 750 species described from various insects. Entomopathogenic fungi were among the first living species applied for biological pest management. Because of their eco-friendliness and bio-persistence, these entomopathogens are recommended to kill insects at various phases of their life cycle (Gul *et al.*, 2014). The most entomopathogenic fungi are found in the Hypocreales (Ascomycota) group, which has a broad host range, and the Entomophthoromycota (Zygomycota) group which is more specialised. Entomopathogens such as *Metarhizium anisopliae* and *Beauveria bassiana* have been thoroughly studied due to their pathogenicity to a variety of insects. *Beauveria bassiana* and *Metarhizium anisopliae* were among the first entomopathogenic fungi to be effectively employed for myco-biocontrol of insect pest.

The *Metarhizium* and *Beauveria* genera are two of the most important EPF groups for practical pest management. They have received the greatest attention globally due to the fact that they are commonly found in nature and have a huge number of hosts that their control is highly effective and that they are simple to utilise in the production of commercial products. They are distinguished by the ability to produce free conidia from reproductive structures (conidiophores) that are easily transported by the wind. EPF are found in a variety of habitats and may be found in a wide range of geographic and climatic areas in both cultivated and natural soils (Vega *et al.*, 2008).

Biology and Life Cycle of EPF

These are related to insects because they utilise chitin as a carbon source for their exoskeleton; their conidia cling to the bodies of the insects and the cycle begins (Fig. 1). Entomopathogenic fungus has a parasitic phase (from host infection to host death) and a saprophytic phase (after host death). Entomopathogenic fungi infect their hosts through the exterior cuticle whereas other entomopathogens (bacteria and viruses) that enter the insects through food. Steps of infection involves: the spore adhering to the insect cuticle, the germ tube penetrating the cuticle, the fungus developing inside the insect body and fungal hyphae colonising the hemocoel. Entomopathogenic fungus spores are frequently coated with a coating of mucus made up of proteins and glucans which aids in their adhesion to the insect cuticle. Several entomopathogenic fungi form specialised structures called appressoria when their spores germinate. The epicuticular surface of the germinating spore is attached by the appressorium. The mechanical pressure and enzymatic activity of the germ tube cause penetration of the insect cuticle. The secretion of successive lipases, proteases and chitinases plays a crucial role in penetration. Most entomopathogenic fungi develop within insects as yeast-like propagules (blastospores), hyphal bodies or protoplasts that lack a cell wall. These structures flow throughout the hemocoel. Mechanical damage induced by developing mycelia inside the insect (mummification) or poisons generated and released by the pathogen are the most common causes of insect death. Toxins having insecticidal activity such as destruxins, oosporein, beauvericin, bassianolide, bassianin, beauveriolides, bassiacridin, cordycepin and ciclosporin have been obtained from EPF & their action and involvement in the disease process has been studied. (Hamil *et al.,* 1969; Susuki *et al.*, 1977; Quesada-Moraga *et al.*, 2006; Wang and Wang, 2017).

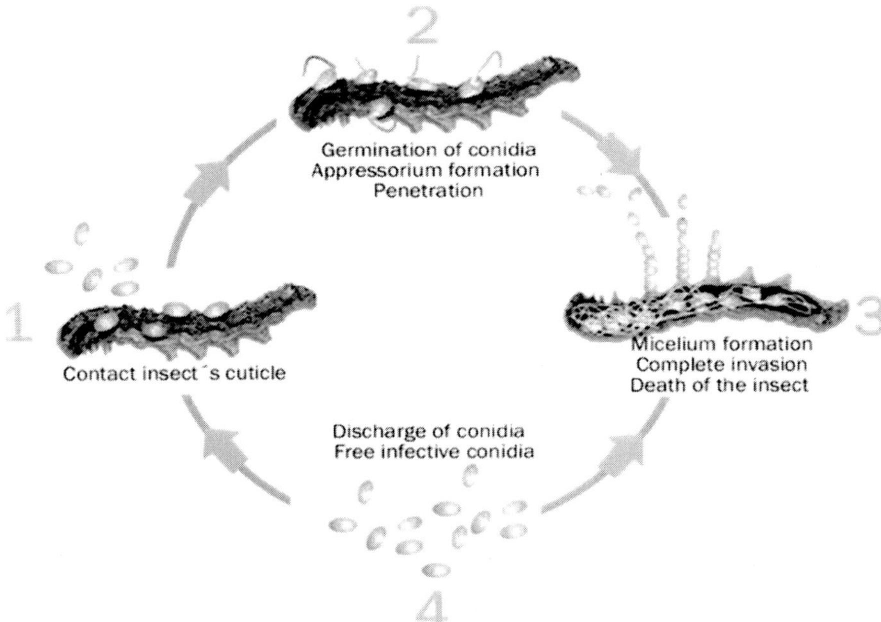

Fig. 1. Life cycle of entomopathogenic fungi. (Barra-Bucarei, *et al.*, 2019)

Isolation of Entomopathogenic Fungi

1. Isolation from Soil

Using the Galleria bait approach based on Zimmermann, the entomopathogenic fungi were isolated from soil samples. Soil samples were taken from 0-20 cm deep in a circle 1 m in diameter around tree trunks. A total of 1.5 kg of soil was taken in plastic bags and transported to the laboratory in a chilled box. Larger wax moth's larvae *G. mellonella* larvae were distributed in the cups at @ of 3 larvae per cup and wetted to about 15 percent moisture content. Infected larvae were moved to 5centimetre diameter Petri plates furnished with wetted filter paper after a week of incubation at 25°C and left for development and sporulation. Both the quantity of infected *G. mellonella* larvae in each sample as well as the quantity of positive samples (samples containing entomopathogenic fungus) were counted.

Isolation of EPF from the Infected Insects

Infected insects covering in fungal mycelium were collected in the field and transported to the laboratory. Different stages of *Cassidavittata* and *Scrobipalpaocellatella* have been identified dead and completely covered by fungal mycelium. The insect cadavers were placed on moist filter paper in a Petri plate and incubated for a week at 24 ±1°C. The new fungal generation was isolated from the surface of an insect carcass and grown on Petri plates using PDA media. Weekly purifications of fungal cultures were done until pure cultures were obtained.

Mass production of Entomopathogenic Fungi

Fungi were grown and maintained on the following medium.

Peptone Medium

This medium contains 10 g peptone, 40 g dextrose, 2 g yeast extract, 15 g agar, 500 mL chloramphenicol and made-up total volume one litre by adding distilled water. The medium was autoclaved at 15 psi and 120°C for 20 minutes and poured into Petri plates (9 cm diameter 1.5 cm). Poured Petri plates inoculated with entomopathogenic fungi and incubate at 25 ±2°C and 85 ±5 R.H. Every 15 to 30 days, the fungal isolates were re-cultured and stored at 4°C.

Potato Dextrose Agar

PDA medium, which contained 200 g peeled potatoes, 20 g dextrose, 20 g agar, and 1000 mL distilled water. PDA media was autoclaved for 20 minutes at 15 psi and 120 °C and poured into Petri plates (9 cm diameter 1.5 cm). Every 14-30 days, the fungal isolates were re-cultured and stored at 4 °C. To restore the isolate's virulence, they were transferred through their native host or the wax moth larvae *G. mellonella.*

Rice Grains

Isolates of *Beauveria bassiana* and *Metarhizium anisopliae* were grown on wetted rice grains. Two kilogrammes of wet rice grains were soaked in boiling water for ten minutes and placed in thermal bags. These bags were autoclaved for 20 minutes at 15 psi and 120 °C, then inoculated with isolates and cultured at 25 ±2 °C for 15 days. Conidia were extracted using distilled water and filtered through cheese cloth to remove mycelium clumps.

Application technology of EPF for insect pest management

The techniques of application of an entomopathogenic fungus influences the treatment's efficacy. Several aspects must be considered, including spore dispersion and coverage to be sure that the pest comes into touch with the fungus as well as environmental conditions that promote the survival and germination of fungus before, during and after application. Dipping the plant or roots, spraying the foliage, treating the soil and passive transmission by vectors are the four basic techniques of fungal application.

1. Dipping

One of the probable applications is dip plant parts or roots in spore suspensions of entomopathogenic fungi while this approach is not widely used, it has gained popularity in recent years for the control of western flower thrips, *Frankliniella occidentalis*. The fungus is assumed to target the larval stage in the soil, However, there is a lack of research proving the efficacy of this technique.

2. Foliar spray

Foliar sprays have historically been the most common method of applying fungus most likely because farmers are familiar with this technique by its use to chemical pesticides. Extensive research has been conducted to maximise fungal efficacy by spraying the vegetation. Fungi from the ground can be applied using a wide range of applicators. Air application has also been used to treat huge areas in Africa to combat locust and grasshopper infestations. It is important to choose the appropriate sprayer for the agricultural field as well as the appropriate formulation for the circumstances. For example, in greenhouses, it is usual practise to apply fungal sprays in the late afternoon to capitalise on the greater relative humidity that occurs at that timewhich facilitates fungal germination.

Ensuring that fungal spores are evenly distributed to the place where the target insect occurs is arguably the most necessary aspect impacting effective control although it is difficult to attain. Kim *et al.*, (2010a) described using supernatant from a *Beauveria bassiana* liquid culture which included a thermotolerant chitinase and an enzyme associated to pathogenicity to improve the efficacy of a fungal spray against aphids. The supernatant fraction was combined with 0.01% polyoxyethylene-(3)-isotridecyl ether (TDE-3) as a spreading agent and integrated into the fungal preparation at 0.5% (w/v) with Attagel (an inert powder used as a thickener developed by BSAF Corp.). This is only one of several ways to improve fungal efficacy using enhanced application and formulation methods.

3. Soil treatments

Soil treatment with fungal spores is possibly the most sensible application approach but it is only appropriate if the target insect has a sensitive soil phase.

In general, natural soil water content levels are adequate to encourage conidial germination and mycelial growth allowing fungal inoculum to be sustained throughout time. Many agricultural settings waters their crops on a regular basis to ensure optimal moisture levels for the growth of fungi. Kim *et al.*, (2010b) evaluated the downward migration of *Beauveria bassiana* in the potting medium with varying amounts of wetness and top-watering due to concerns about spores being leached out during watering. They discovered that the concentration of *Beauveria bassiana* was highest at the top 3 cm of the potting medium and that there was no crucial movement of the spores after 18 days of frequent watering.

Another advantage of introducing fungus to soil rather than foliage is that the conidia are shielded from UV radiation damage within the soil. Temperature also has a significant impact on fungal effectiveness. Soil helps to moderate big temperature swings which generally encourage fungus growth and infectivity. The presence of antagonistic fungi, bacterial or fungal-feeding arthropods and microorganisms is a key potential adverse side effect of treating soil with fungi.

3. Indirect Transmission by Vector

Entomopathogenic fungi are easily transmitted in nature by non-pest species to pest-infested areas. For example, Collembolans which are found in soil are known to carry fungal conidia that adhere to their cuticle. According to research, certain species were able to transport enough fungal propagules to kill mealworms (Dromph, 2003). Other natural cases of indirect vectoring have been described (Baverstock *et al.*, 2010) but the impact of this random transmission on pest populations is unknown. Kapongo *et al.*, (2008) evaluated the ability of utilising bumble bees (*Bombus impatiens*) a popular pollinator in greenhouse-grown vegetables, to transfer *Beauveria bassiana* to greenhouse whitefly and tarnished plant bug, *Lygus lineolaris*, in caged greenhouse experiments. Following fungal dissemination by bees, death rates of up to 56% for whiteflies and 67% for the tarnished plant bug were reported depending on the spore concentration observed. The bees exposed to the fungus died, indicating the delicate balance between positive and negative effects that must be established for a system like this to succeed.

Several commercially available formulations of entomofungal pathogens for crop pest control have been produced. Table 01 depicted a list of commercially available fungi, crops, and target insect pests.

Table 1. This table shown major ENF, crops and their targeted pests.

S. No.	Entomopathogenic Fungi	Crops	Targeted pests
1.	*Aeschersorzia aleyrodis*	Cucumber, potato	*Bemisia tabaci*
2.	*Beauveria bassiana*	Cotton	*Aphis gossypii, Chortoicetes terminifera, Rachiplusia nu*
		Tomato	*Helicoverpa armigera, Helicoverpa zea,*
		Coffee	*Hypothenemus hampei,*
		Sorghum	*Chilo partellus, Sesamia nonagrioides*
		Maize	*Ostrinia nubilalis, Sesamia calamistis*
		Cauliflower	*Bemisia tabaci*
3.	*Beauveria brongniartii*	Sugarcane	*Tomarus subtropicus*
4.	*Verticillium lecanii*	Glasshouse crops	*Aphids, Whiteflies and Thrips*
5.	*Metarhizium anisopliae*	Fava bean	*Acyrthosiphon pisum*
		Rapeseed	*Aphis fabae, Plutella xylostella*
		Pepper	*Myzus persicae*
		Strawberry	*Myzus persicae*
6.	*Metarhizium brunneum*	Soybean	*Aphis glycines,*
		Melon	*Aphis gossypii*
		Cauliflower	*Bemisia tabaci*
7.	*Lecanicillium muscarium*	Cauliflower	*Plutella xylostella*
8.	*Lecanicillium lecanii*	Cotton	*Aphis gossypii*
9.	*Purpureocillium lilacinum*	Cotton	*Aphis gossypii*
10.	*Clonostachys rosea*	Coffee	*Hypothenemus hampei*
11.	*Aspergillus parasiticus*	Cotton	*Chortoicetes terminifera*
12.	*Isaria fumosorosea*	Sorghum	*Sesamia nonagrioides*

Sources. Sinha *et al.,* (2016), Mantzoukas *et al.*, (2020)

Advantages

- They are only harmful to insects/arthropods and not plants or animals. They do not leave toxic residues on crops as some chemical pesticides do.
- They are somewhat host-specific allowing them to fit into an IPM programme that includes other beneficials such as predators and parasitoids.

- The majority are considered safe for humans and pose no disposal threat especially when compared to chemical insecticides (Laird *et al.*, 1990).
- Compared to many parasitoids and predators, entomopathogenic fungi may be mass-produced on a solid substrate in huge quantities at a cheap cost and they can be maintained with little loss of viability or efficacy.
- The majority of entomopathogenic fungi can endure a wide range of temperatures (often 0-40°C).

Disadvantages

- More time consuming than chemical pesticides, for example, fungus take 2-3 weeks to kill the insects. In comparison, this chemical pesticide may do the same thing in a short time as 2-3 hours.
- Due to the high specificity of the fungal pesticide additional control methods are necessary for other insects.
- The production cost of bioinsecticides is substantially greater with a shorter shelf life and the requirement of cold storage for spore maintenance.
- The use of the biocontrol agent must coincide with contributing conditions such as increased humidity, a low number of pests and a fungicidal-free environment.
- Poses a danger to those with immunological deficiencies or who are immunocompromised.
- Because persistence and efficacy fluctuate amongst insect species regular optimisation of approaches is essential to maintain long-term impacts.

Conclusion

One of the most promising agents for the biological control of insects is entomopathogenic fungi. Entomopathogenic fungi are a group of fungi that target and infect their insect hosts to kill them (Singkaravanit*et al.*, 2010). The Entomophthoromycota (Zygomycota) which are more specialised and the Hypocreales (Ascomycota) group which has a wide variety of hosts have the most entomopathogenic fungus species. Entomopathogenic fungi go through a parasitic phase (from host infection to host death) and a saprophytic phase (after host death) throughout their life cycle. Utilising the Zimmermann-based Galleria bait approach to isolate the entomopathogenic fungus from soil samples it may be more significant to focus on mass-producing EPF at cheap cost and in huge quantities. Growers rely on the commercial availability of substances made from fungi and those that are reasonably simple to make have a higher likelihood of being sold. The majority of commercially available fungus are *Beauveria, Metarhizium, Lecanicillium* and *Isaria species* since they are all reasonably simple to mass manufacture. For usage against agricultural insect pests, fungi such as *Beauveria bassiana, Vuillemin, Isaria fumosorosea* and *Metarhizium anisopliae* are currently being tested. Entomopathogenic fungi will soon play a significant role in the treatment of insect pests. But in terms of effectiveness and acceptance, it still lags far below synthetic chemicals. Understanding their further functions in nature as endophytes, competitors of pathogens that are associated with the rhizosphere and even plant growth-promoting factors requires further study. As a result, the EPF deserves the best possible prospects in the advancement of agriculture because of its various responsibilities.

References

Abdel-Raheem, M.A. (2005). Possibility of using the Entomopathogenic Fungi Beauveria bassiana and Metarhizium anisopliae for controlling the sugar-beet insects Cassidavittata Vill. and Scrobipalpa ocellatella Boh. Egypt, Ph.D. Faculty of Agriculture, Cairo University, Cairo, 86

Araújo, J.P., and Hughes, D.P. (2016). Diversity of entomopathogenic fungi: which groups conquered the insect body. Advances in Genetics, 94: 1-39

Barra-Bucarei, L., France Iglesias, A. and Pino Torres, C. (2019). Entomopathogenic Fungi. Natural Enemies of Insect Pests in Neotropical Agroecosystems: Biological Control and Functional Biodiversity, 123-136.

Baverstock, J., Roy, H.E. and Pell, J.K. (2010). Entomopathogenic fungi and insect behaviour: from unsuspecting hosts to targeted vectors. The Ecology of Fungal Entomopathogens, 89-102

Culliney, T.W. (2014). Crop losses to arthropods. Integrated Pest Management: Pesticide Problems, 3: 201-225

Dromph, K.M. (2003). Collembolans as vectors of entomopathogenic fungi. Pedobiologia, 47(3): 245-256

El-Wakeil, N., Saleh, M., and Abu-hashim Mohamed (Eds.). (2020). Cottage Industry of Biocontrol Agents and Their Applications, Springer Nature Switzerland. doi:10.1007/978-3-030-33161-0

Gul, H.T., Saeed, S. and Khan, F.A. (2014). Entomopathogenic fungi as effective insect pest management tactic: A review. Appl. Sci. Bus. Econ, 1(1): 10-18

Hamill, R.L., Higgens, C.E., Boaz, H.E., and Gorman, M. (1969). The structure op beauvericin, a new depsipeptide antibiotic toxic to Artemia salina. Tetrahedron Letters, 10(49): 4255-4258

Hoffmann, M. and Frodsham, A. (1993) Los Enemigos Naturales de las Plagas de Insectos Vegetales, Ithaca, New York: In Extension Cooparativa de la Universidad de Cornell

Humber, R.A. (1996). Fungal pathogens of the Chrysomelidae and prospects for their use in biological control. Biology of Chrysomelidae, 3: 1-40

Kapongo, J.P., Shipp, L., Kevan, P., and Broadbent, B. (2008). Optimal concentration of Beauveria bassiana vectored by bumble bees in relation to pest and bee mortality in greenhouse tomato and sweet pepper. BioControl, 53: 797-812

Kidanu, S. and Hagos, L. (2020). Entomopathogenic fungi as a biological pest management option: A review. Int. J. Res. Stud. Agric. Sci, 6: 1-10

Kidanu, S. and Hagos, L. (2020). Research and application of entomopathogenic fungi as pest management option: a review. Journal of Environment and Earth Science, 10(3): 31-39

Kim, J.S., Roh, J.Y., Choi, J.Y., Wang, Y., Shim, H.J., and Je, Y.H. (2010a). Correlation of the aphicidal activity of Beauveria bassiana SFB-205 supernatant with enzymes. Fungal Biology, 114(1): 120-128

Kim, J.S., Skinner, M., Gouli, S., and Parker, B.L. (2010b). Influence of top-watering on the movement of Beauveria bassiana, GHA (Deuteromycota: Hyphomycetes) in potting medium. Crop Protection, 29(6): 631-634

Kumar, D. and Kalita, P. (2017). Reducing postharvest losses during storage of grain crops to strengthen food security in developing countries. Foods, 6(1): 8

Lacey (1997). Manual of techniques in insect pathology. Academic Press (Elsevier), NY, p. 409

Laird, M., Lacey, L.A. and Davidson, E.W. (1990). Safety of microbial insecticides. CRC Press Inc

Mantzoukas, S., and Eliopoulos, P. A. (2020). Endophytic entomopathogenic fungi: A valuable biological control tool against plant pests. Applied Sciences, 10(1): 360

Quesada-Moraga, E.E.A.A., Maranhao, E.A.A., Valverde-García, P. and Santiago-Álvarez, C. (2006). Selection of Beauveria bassiana isolates for control of the whiteflies Bemisia tabaci and Trialeurodes vaporariorum on the basis of their virulence, thermal requirements, and toxicogenic activity. Biological Control, 36(3): 274-287

Roberts (1980). Fungi toxins. In: Bureges HD (ed)Microbial control of insects, mites, and plant diseases, vol 2. Academic Press, New York, pp. 223-248

Ruiu, L. (2015). Insect pathogenic bacteria in integrated pest management. Insects, 6(2): 352-367

Singkaravanit, S., Kinoshita, H., Ihara, F. and Nihira, T. (2010). Cloning and functional analysis of the second geranylgeranyl diphosphate synthase gene influencing helvolic acid biosynthesis in Metarhizium anisopliae. Applied Microbiology and Biotechnology, 87: 1077-1088

Sinha, K.K., Choudhary, A.K. and Kumari, P. (2016). Entomopathogenic fungi. In Ecofriendly pest management for food security (pp. 475-505). Academic Press

Skinner, M., Parker, B.L., and Kim, J.S. (2014). Role of entomopathogenic fungi in integrated pest management. Integrated Pest Management, 169-191

Souza, B., Vazquez, L.L., and Marucci, R.C. (2019). Natural enemies of insect pests in neotropical agroecosystems. Biological control and functional biodiversity. Springer, Cham, 1: 546

Suzuki, A., Kanaoka, M., Isogai, A., Tamura, S., Murakoshi, S. and Ichinoe, M. (1977). Bassianolide, a new insecticidal cyclodepsipeptide from Beauveria bassiana and Verticillium lecanii. Tetrahedron Letters, 18(25): 2167-2170

Vega, F.E., Posada, F., Aime, M.C., Pava-Ripoll, M., Infante, F., and Rehner, S.A. (2008). Entomopathogenic fungal endophytes. Biological control, 46(1): 72-82

Wang, C. and Wang, S. (2017). Insect pathogenic fungi: genomics, molecular interactions, and genetic improvements. Annual Review of Entomology, 62: 73-90

Zimmermann, G. (1986). The 'Galleria bait method' for detection of entomo-pathogenic fungi in soil. Journal of Applied Entomology, 102(1-5): 213-215

11

Chemical Communication Between Plants and Natural Enemies (Carnivorous Insects and Mites)

Sonkamble M.M., Matre Y.B*. and P.S. Neharkar

Department of Agricultural Entomology, Vasantrao Naik Marathwada Krishi Vidyapeeth, Parbhani, Maharashtra

Introduction

Chemical communication between plants and insects is an essential part of mutual relationships, in which both partners benefit from the communication. The role of volatiles in pollination systems provides an excellent model for understanding the ecology and evolution of chemical communication between plants and insects. Infochemical:- A chemical that conveys information in an interaction between two individuals evoking in the receiver a behaviour or physiological response. Chemicals involved in interspecific communications were given the name of Allelochemicals by R. H. Whittaker in 1970. Allelochemicals affect the survival, growth & development of herbivorous insects as well as their natural enemies. Plant produced substances that serves as attractants, arrestants, oviposition and feeding stimulants for herbivores are the best known kairomones. These plant derived substances also influence organisms of the third trophic level when they are requested by herbivores & used by natural enemies to locate the herbivore. Communication can be active or passive. For e.g. a beetle may actively defend itself by ejecting a noxious spray when attacked or a caterpillar may leave secretions on a leaf while feeding that provide a cue for a parasitoid searching for the caterpillar. Chemical communications play a critical role in life for survival & for finding food & mates. An insect produce strongly repellent chemicals against a predator or parasitoid & actively reject them at the time of attack. Chemical ecology is the science that addresses the role of chemical cues in the interaction of organisms with their environment. Example of chemical communication is the use of sex pheromones by insects such as the silk moth *Bombyx mori.*

What is tritrophic interaction

An interaction among host plants, herbivores and their natural enemies (parasitoids & predators) is called tritrophic interaction. It leads to effective defence & attack at each level.

Two types of defence

1. **Intrinsic defence:** When the plant alone produces defence through physical means or through production of chemicals or both it is called intrinsic defence.
2. **Extrinsic defence:** When the natural enemies of insect pests benefit the host plants by reducing the pest abundance is called extrinsic defence.

1. **Intrinsic defence**

- The physical, morphological, physiological and chemical characteristics of plants interact with the parasitoids and predators of insects by influencing their host seeking ability and affecting the efficacy with which they locate and utilize hosts.
- Plants provide nutrition to the natural enemies in the form of pollen and nectar.
- Insects such as aphids and whiteflies which secrete honey dew attract ants which also keep herbivores away.
- Plants that promote the effectiveness of the carnivores enemies of their own enemies & in this way the effect of the herbivore. The bitrophic defence is referred to as direct or intrinsic defence & the defence that employes carnivores is called indirect or extrinsic defence.

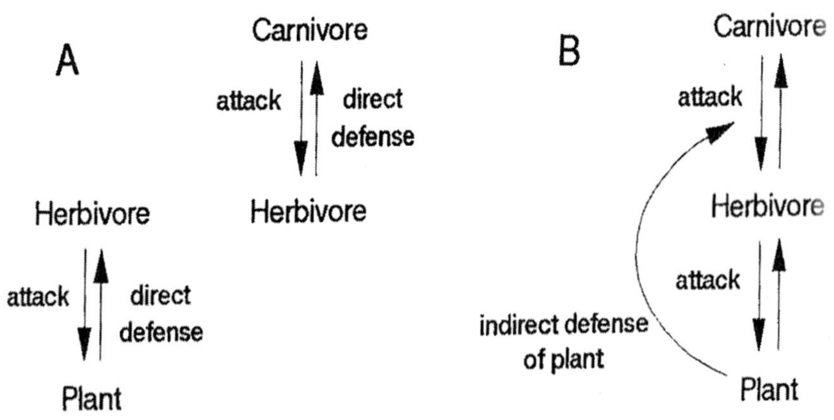

Attack-defense interactions in (A) a bitrophic context and (B) in a tritrophic context

Fig. 1. Attack – defense interactions

Tri-trophic interactions: As well-known, plants recruit natural enemies to protect them from herbivores through herbivore-induced plant volatiles when plants are damaged by the herbivores. Natural enemies have to utilize different info chemicals to locate the appropriate host plants and host prey in their whole life span.

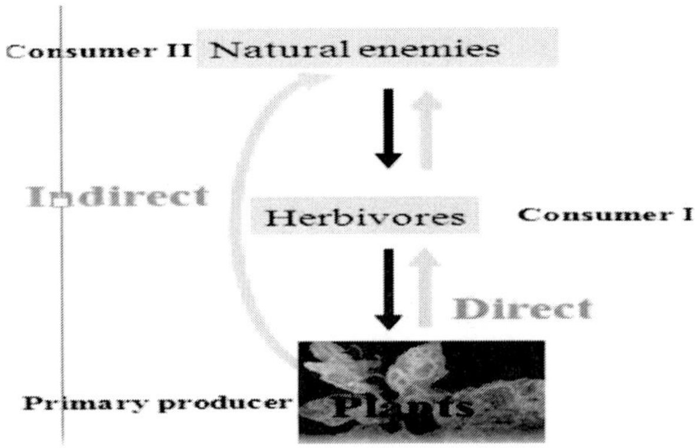

Fig. 2. Tri-trophic interaction

Plant volatiles play a critical role as signals in tri-trophic level interactions. These are interactions involving three trophic levels; for example, plants (1st trophic level), herbivores (2nd trophic level), and the natural enemies of herbivores (predators and parasitoids) (3rd trophic level). Plants damaged by herbivores often produce a blend of volatiles commonly referred to as herbivore-induced plant volatiles (HIPVs) .The release of HIPVs may signal the presence of potential prey or hosts and, therefore, can be exploited by natural enemies to locate the prey organism (Sabelis *et al.*, 1999; Verkerk, 2004). Herbivore-induced plant volatiles (HIPV) play an important role in host/prey searching behavior of predators and parasitoids of herbivorous arthropods (Heil, 2008; Dicke and Baldwin, 2010).

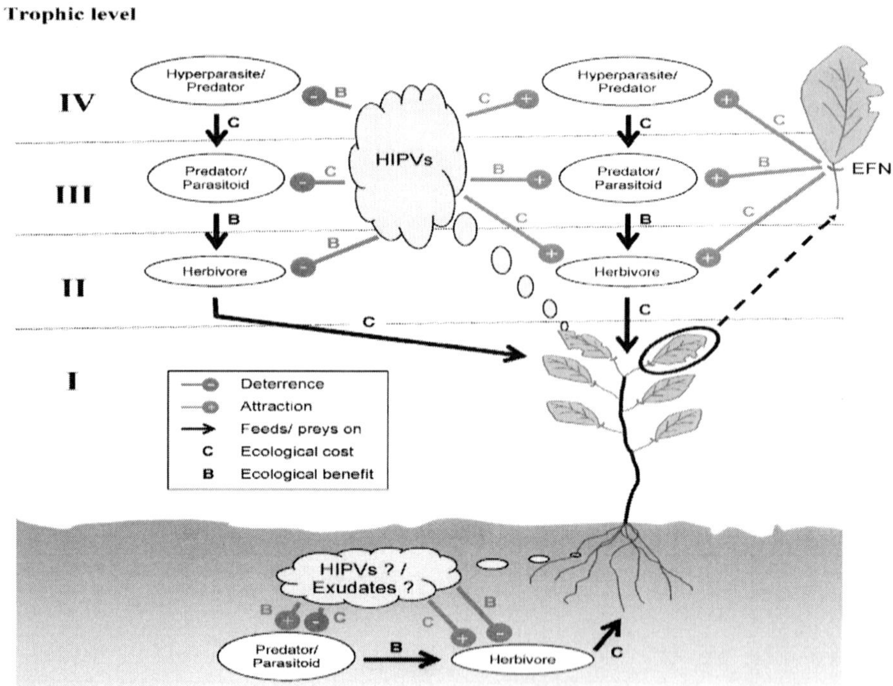

Fig. 3. Potential interactions of plants with higher trophic levels via herbivore-induced plant volatiles (HIPVs) and extrafloral nectar (EFN).

Bark beetles (Curculionidae: Scolytinae) provide a useful system for exploring ecological and evolutionary implications of semiochemical interactions among predators and prey. They spend all but a brief portion of their life cycle within the subcortical tissues of host plants, and rely on chemical signaling for a variety of life history and ecological processes, including mating, host plant procurement, and niche partitioning (Wood 1982). Methyl salicylate is a known plant defence inducer and has been used successfully in the field to protect cereal crops against pest aphids (Prinsloo *et al.*, 2007). Several different plant species, including corn *(Zea mays)* infested by caterpillars of the beet army worm *Spodoptera exigua,* take revenge on this pest by attracting the parasitoid *Cotesia marginiventris.* In lima beans, various coordinated defence strategies against mite infestation have been discovered. First, they change their scent to make them unattractive to the mites. Then the plants emit scents that are perceived by other plants, which then do precisely the same thing to warn surrounding lima beans before the mites even reach them. Some of the emitted substances have the effect of attracting other mites that eat the attacking red mites .Similar defence processes have been described in tomato plants (Kant *et al.*, 2004, Pearce and Ryan, 2003). *Tetranychus urticae* Koch infestation or

JA treatment induced the emission of volatiles in cucumber plants that attract the predatory mite *Phytoseiulus persimilis.* In Lima bean plants, infestation by the spider mite *Tetranychus urticae* leads to the production of volatile terpenoid chemicals such as (E)-P.ocimene, 4, 8-dimethyl-1,3(E), 7-nonatriene and linalool as well as methyl salicylate. The volatile compounds produced by the plants have been found to attract the predato mite *Phytoseilus persimilis*.

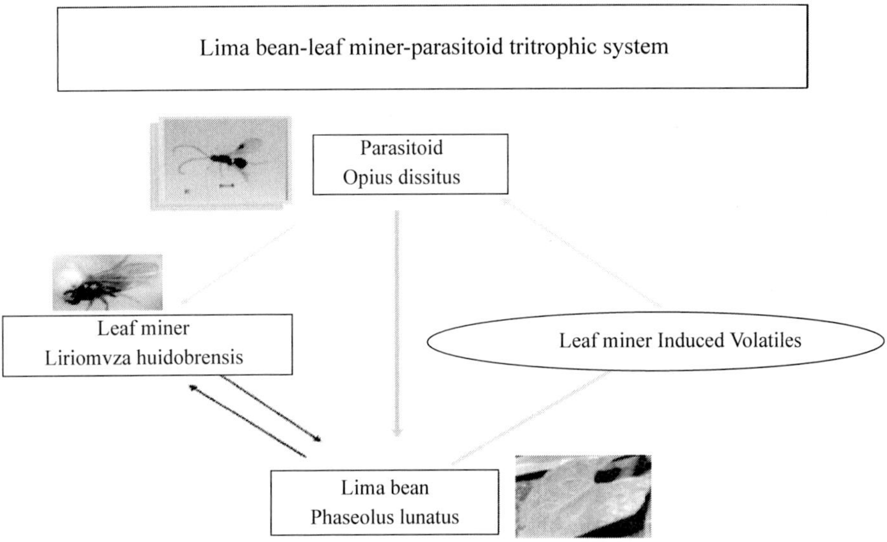

Fig. 4. Lima bean-leafminer-parasitoid tritrophic system

Jasmonic acid (JA) and its volatile derivative methyl jasmonate (MeJA) are phytohormones involved in plant defenses against herbivores (Karban & Baldwin, 1997), Salicylic acid is a phytohormone often associated with plant resistance against pathogens but also against sucking insects such as aphids and whiteflies (Walling, 2000) 'Activation of both the JA and SA pathways by *T. urticae* is required to attract predatory mites to damaged lima bean plants (Dicke *et al.*, 1999). Rodriguez- Saona *et al.* (2011a) used MeSA to attract natural enemies in agricultural fields and found that natural enemies (i.e.,coccinellids, syrphids, lacewings, predatory bugs, and parasitic Hymenoptera) are broadly attracted to MeSA. MeSA is now commercially available as PredaLure® to attract natural enemies of agricultural insect pests.

The quantity of VOCs released by damaged plants is much larger than the amount of VOCs released by the actual herbivores, for example herbivorous mites and insects1 or by their faeces. the specific VOC signals released by plants after damage by a herbivore are important signals that improve the fitness of the plant by eliciting behavioural responses in herbivore natural enemies and thus increasing the predation rate leading to reduced plant damage.

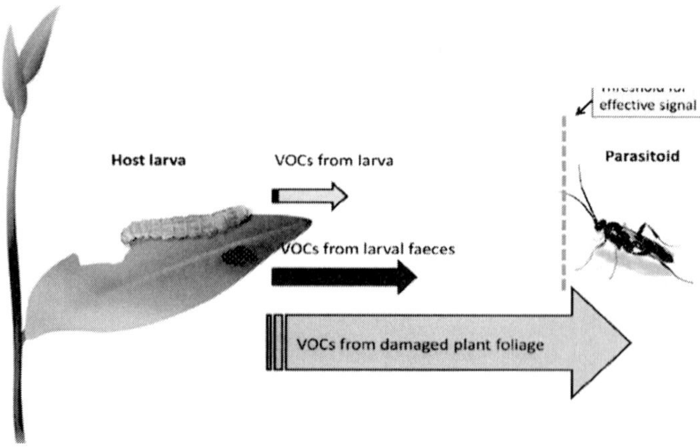

Fig. 5. Schematic illustration of interactions between herbivores and their natural enemies (parasitoids communicated via plant volatile molecules).

Ant plants are prime examples of a plant-predator symbiotic mutualism. Plants provide resources, largely constitutively produced to ant species that may be specialized to nest and feed on these plants. The ants consume extra floral nectar (EFN) and food bodies (FBs). The production of ant food rewards increases predator visitations, and thus predation on herbivores may be higher on EFN-producing plants than on plants not producing EFN (Heil & McKey 2003).

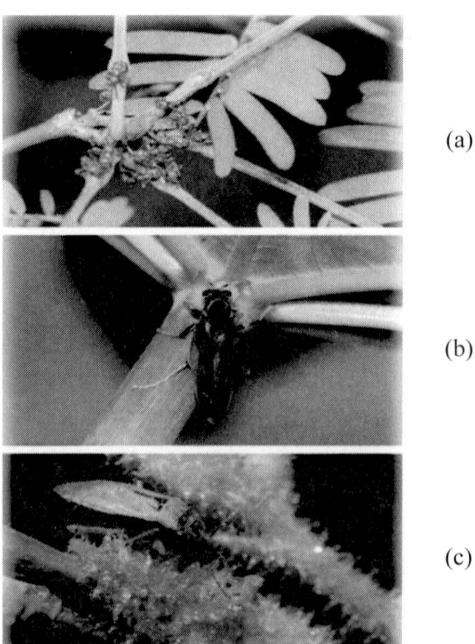

Fig. 6. Resource-mediated and information-mediated indirect resistance.

a) Obligate ant-ant plant associations
b) Facultative attraction of generalist predators with extrafloral nectar
c) Herbivory-induced volatile organic compound (HIVOC) emission as predator attracting cue

Volatile interactions between undamaged plants affect higher trophic levels

The natural enemies of herbivores respond to habitat cues denoting prey abundance and quality. Specialised natural enemies, such as parasitic wasps, use cues indicating the presence of their hosts whereas polyphagous predators respond to more general indicators of habitat quality (Pettersson *et al.* 2005). Johnson (2008) found that plant genotype diversity can directly affect the abundance of insect predators in a natural ecosystem.

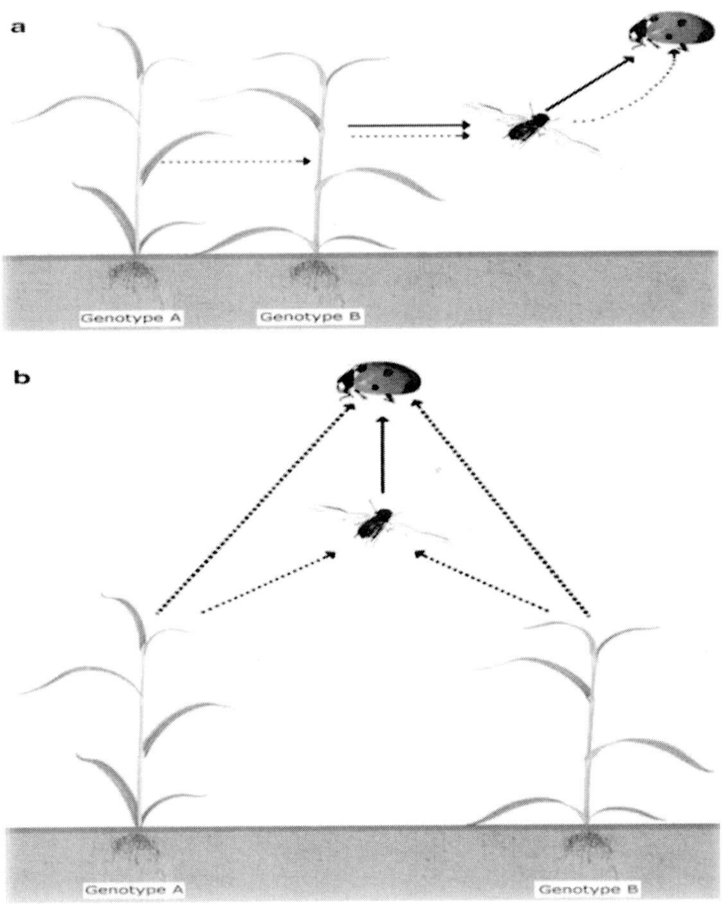

Fig. 7 a & b. Two mechanisms by which mixing different plant genotypes may affect the natural enemies of herbivores via volatile chemicals (dotted lines) and host quality (bold lines).

In (a) volatiles from one plant genotype induce responses in a second genotype that affect plant volatile emission and herbivore host quality. In (b) mixed odours of the different plant genotypes affect natural enemy orientation

Predator-released semiochemicals and tritrophic interactions

In tritrophic interactions the semiochemicals released by natural enemies cannot be neglected because these also alter the behavior or physiology of plants or herbivores. Some predators release synomones which are acquired from plants, as many of the predators and parasitoids obtain food from plants. Plants respond to synomones produced by predators. predators may also produce allomones to attract their prey, or to camouflage their own true identity. Carnivore populations are expected to vary in their responses to plant volatiles & other major foraging cues. The conspiracy between plant & carnivore is based on the feeding activity of the herbivore itself. Escape from carnivores can be a driving force in the evolution of plant selection by herbivores. The chances of a herbivore successfully escaping from carnivores through plant selection are best when there is a relative focus of the carnivores on plant stimuli over herbivore stimuli i.e carnivores specialized at the plant level (Vet& Dicke 1992).

Synergism in tritrophic interactions

Synergism plays an important role in insect-plant interactions as well as tritrophic interactions. Number of plants damaged by insects emit chemical signals in large amounts that guide natural enemies to the concerned insects, The mites *Tetranychus urticae* & *Panonychus ulmi* are known to affect several host plants, including apple leaves & release volatiles after feeding, eliciting different responses from predatory mites. In tritrophic interactions, herbivore-induced, carnivore attractants play an important role. Herbivore –induced plant volatiles provide natural enemies with information about plant species, plant cultivars & the herbivore insect species infesting the plant. Thus the herbivore-induced volatiles are essential information source for natural enemies & act as communication signals. Plants communicate with the members of the third trophic level, since plants under attack by insects emit chemical signals that guide natural enemies to the site of plant damage. Semiochemicals mediating host finding emanate from frass, mandibular, labial & other secretions, exuviae and moth scales & damage tissues associated with the plant.

Conclusions

1. Chemical communication between plants and insects is an essential part of mutual relationships, in which both partners benefit from the communication. For example.

- Methyl benzoate is the main component in the leaf odours of *Sarracenia leucophylla* and *S. flava* is a typical component from flowers of bee and moth pollinated plants.
- In the essential oil of *S. flava* leaf extract 2- phenyl ethanol is found which helps in attracting insect species from Coleoptera, Diptera, Hymenoptera and Lepidoptera.

2 Herbivore induced plant volatile play an important role in host/ plant searching behavior of predators and parasitoids of herbivorous arthropods. For example

- Jasmonic acid (JA) treatment induced the emission of volatiles in cucumber plants that attracts the predatory mites, *Phytoseilus permicilis.*
- Methyl salicylate (MeSA) used to attract natural enemies in agricultural fields and found that natural enemies (Coccinellids, Syrphid, Lacewings, Predatory bugs and Parasitic Hymenoptera) are broadly attracted to MeSA.
- Tomato plants by spraying with jasmonic acid attract parasitoid wasps that prey upon tomato caterpillar

3. EFNs are specialized nectar secreting organs that may occur in all vegetative and reproductive plant structures. For example
   - Besides ants, EFNs attract Coleoptera, Diptera, Hymenoptera .
   - Both ants and wasps exerted beneficial effects on the EFN bearing plant *Turnera ulmifolia.*
4. Plant volatiles play a critical role as signals in tritrophic level interactions. For example
   - Lady bird beetle *Coccinella septumpunctata* make extensive use of volatile chemical cues in locating habitats, plants and prey.
   - The volatile compounds produced by Lima bean, *Phaseolus lunatus* have been found to attract the Predatory mite, *Phytoseilus persimilis.*
5. The quantity of VOCs released by damaged plant is much larger than the amount of VOCs released by the actual herbivores. For Example
   - VOC emission profiles of healthy intact cabbage plants and Diamond Back Moth (*Plutella xylostella)* larvae-damaged cabbage plants.
6. Synergism plays an important role in insect-plant interactions as well as tritrophic interactions. For Example
   - An enhanced phagostimulatory effect is seen when sinigrin and glucose interact as evident in *Plutella xylostella.*

- Piperonyl butoxide is known to be an effective synergist in conjunction with diflubenzuron in the case of *Spodoptera exigua* increasing the toxicity of diflubenzuron.

Future line of study

1. Need to study whether chemical communication is more in natural or managed ecosystems.
2. Need to reveal fascinating new aspects of the ecology of plant communications.
3. Exogenous elicitors may be developed synthetically and applied to a crop and utilized to increase time of searching by the natural enemies.
4. Very little is known about the interaction of plants and insects under tropical conditions.
5. Future attempts should be made to understand why plant species differ in the way they attract carnivores and under what circumstances they do so.
6. In order to fully appreciate the influence of chemical signaling on community and ecosystem processes one needs thorough knowledge of the mechanism of chemical information nveyance from the gene to individual.

References

Dicke, M. (1999). Are herbivore-induced plant volatiles reliable indicators of herbivore identity to foraging carnivorous arthropods? Entomol. Exp. Appl. 91: 131-142

Dicke, M., and Baldwin, I.T. (2010). The evolutionary context for herbivore-induced plant volatiles: beyond the 'cry for help'. Trends Pl. Sci. 15: 167-175

Heil, M. and McKey, D. (2003) Protective ant-plant interactions as model systems in ecological and evolutionary research. Annual Review of Ecology, Evolution, and Systematics, 34: 425-453

Heil, M. (2008). Indirect defence via tritrophic interactions. New Phytol. 178(1): 41–61. Isolation and characterization of three novel defense-signalling glycopeptide hormones coded in a single precursor gene. J. of Bio. Chem., 278(32): 30044-30050

Johnson M.T.A. (2008). Bottom-up effects of plant genotype on aphids, ants, and predators. Ecology 89: 145-154

Kant, M.R., Ament K., Sabelis M.W., Haring M.A. and Schuurink R.C. (2004). Differential timing of Spider Mite-Induced Direct and Indirect Defenses in Tomato Plants. Pl. Physiol. 135: 483-495

Karban, R. & Baldwin I.T. (1997). Induced responses to herbivory. The University of Chicago Press

Pearce G. and Ryan, C. (2003). Systemic signaling in tomato plants for defense against herbivores

Pettersson, J., Ninkovic, V., Glinwood, R., Birkett, M.A. and Pickett, J.A. (2005) Foraging in a complex environment – semiochemicals support searching behaviour of the seven spot ladybird. Eur J Entomol 102: 365–370

Prinsloo, G., Ninkovic, V., Linde, T.C.V.D., Westhuizen, A.J.V.D., Pettersson, J. and Glinwood, R. (2007). Test of semiochemicals and a resistant wheat variety for Russian wheat aphid management in South Africa. J. Appl. Entomol., 131: 637-644

Rodriguez-Saona, C., Kaplan, I., Braasch, J., Chinnasamy, D. and Williams, L. (2011a). Field responses of predaceous arthropods to methyl salicylate: a meta-analysis and case study in cranberries. Biol. Control, 59: 294-303

Sabelis, M.W., Janssen, A., Pallini, A., Venzon, M., Bruin, J., Drukker, B. and Scutareanu, P. (1999). Behavioral responses of predatory and herbivorous arthropods to induced plant volatiles: from evolutionary ecology to agricultural applications, In: Induced Plant Defenses Against Pathogens and Herbivores, eds. A.A. Agrawal, S. Tuzun & E. Bent, pp. 269-296, APS

Verkerk, R.H.J. (2004). Manipulation of tritrophic interactions for IPM, In: Intregrated Pest Management: Potential, Constraints and Challenges, eds. O. Koul, G.S. Dhaliwal & G.W. Cuperus, 55-71

Vet, L.E.M. and Dicke, M. (1992). Ecology of infochemical use by natural enemies in a tritrophic context. Ann. Rev. of Entomol., 37(1): 141-172

Walling, L.L. (2000). The myriad plant responses to herbivores. J. of Pl. Growth Regulation, 19: 195-216

Wood D. (1982). The role of pheromones, kairomones, and allomones in the host selection and colonization behavior of bark beetles. Ann Rev Entomol 27: 411–446

12

Role of Nuclear Polyhedrosis Virus in Insect Management

Mateti Gayithri

Department of Plant Pathology, Assam Agricultural University, Jorhat, Assam

Introduction

In most developing nations, the primary method of managing insect pests on various crops is by synthetic insecticides. One of the fastest and most efficient ways to reduce insect populations is by chemical control where a farmer quickly achieves extraordinary results.

However, overusing pesticides indiscriminately and without proper research for longer periods of time led to a number of issues, including the risk of environmental contamination, loss of biodiversity, development of insecticide-resistant pest populations, resurgence, outbreaks of secondary pests, increase in chemical inputs, toxicological hazards due to accumulation of pesticide residues in the food chain, etc. These issues ultimately contribute not only to inefficient insect control but also environmental and health hazards.

The use of chemical pesticides for the management of insect pests must there fore be rationalised immediately. The preference for entomopathogenic Nuclear Polyhedrosis Virus (NPV) among the different biological control agents employed as an alternative to synthetic insecticides has been increasing in recent years. The normal physiological functions of the targeted species are also disrupted by novel insecticides that have been developed from natural sources by considering insect development mechanisms like moulting as well as chloride channel activator. The public's rising concern regarding the potential health risks of synthetic pesticides has recently prompted researchers to look at alternative pest control methods, such as the use of integrated pest management (IPM). IPM is the most efficient, environmentally responsible, and socially acceptable method of managing pests, diseases, and weeds. It integrates cultural, biological, and chemical methods. Under the integrated pest management (IPM) programme, insect harmful viruses like baculoviruses have been utilised successfully for biological control.

History

In the past Germany was the country that initially introduced NPV to *Lymantria monacha* populations in pine forests in 1892. The first field application of NPV was made against gipsy moths in the United States in 1913. A Californian farmer collected *C.eurytheme* larvae that had been killed by NPV and used them as an inoculum for a lucerne crop. The first viral pesticide (Elcar) used to treat cotton bollworm (*Helicoverpa zea*) was registered in the United States in 1975. Arora and Padmanaban (2002). Because they leave no residue and have no negative effects on beneficial predators and parasites, baculoviruses offer an alternative to chemical insecticides for treating certain insect pests.

More than 65 crops, including maize, wheat, sorghum, different legumes, solanaceous crop, malvaceous crop, etc., are attacked by the Helicoverpa and Heliothis Cunningham (1995). Numerous significant *Helicoverpa* and *Heliothis* species, including *Helicoverpa armigera*, *Helicoverpa assulta*, *Heliothis virescens*, and *Helicoverpa zea*, were discovered to be NPV sensitive.

Structure

The virus that causes nuclear polyhedrosis is a obligate pathogen. The virus has a rod-shaped nucleocapsid that measures between 250 and 400 nm in diameter and 40 and 70 nm in height. The virions, are the proteinaceous polyhedral occlusion bodies that house the nucleocapsids in their envelope. The viruses are released by the alkaline stomach juice and attack the ganglia, brain, adipose bodies, haemocytes, tracheal matrix, and nuclei of cell tissue Harappa (1970).

Infection

The normal way of infection is by ingestion of polyhedra or in some cases virions. Larvae should be susceptible because when polyhedra are consumed along with food, occlusion-causing protinaceous substances begin to disintegrate in an alkaline environment ($pH > 9$). The virions infect the midgut epithelial cells; fresh virions are produced in the nuclei of infected cells and infect hemocoel cells as well as other tissues like the body's fat. In that tissue, virions are enclosed in polyhedra and continue to exist there until cell lysis.

Polyhedra can be seen in the nuclei of infected cells 24 to 72 hours after the first infection. Larvae eventually perish and splatter on soil and foliage after a few days. Predators and parasites then consume or spread to other insects or locations. Infected insects have bland colours, are less active, and the ventral side of the larvae turns reddish pink. The epidermis becomes extremely brittle and eventually disintegrates in advanced stages, and the worms become flaccid Deb *et al.* (2015). On plants, infected larvae hang upside down. "Tree top disease" or "wipfelkrankheit" is the name given to this condition. Grasserie diseases are NPV-related illnesses that affect silkworms.

Mass Production of NPV

The mass production of the NPV is carried out in the early fifth instar stage of the *Spodoptera litura* which yields the maximum amount of the NPV. As a result, a continuous culture of the insects can be maintained in the host culture laboratory using the correct handling techniques. The larvae are raised on a diet held in 5 ml glass vials, and when they reach the required maturity, they are moved to a facility that produces viruses. Feeding the semi-synthetic food covered with a clean inoculum of the previously standardised NPV multiplies the NPV.

This is carried out by placing aliquots of 10 ml of viral suspension of concentration of 1x108 polyhedral occlusion bodies (POB) either in glass vials or directly on the surface of the diet, and spreading the suspension evenly with a polished glass rod. After 15 minutes, larvae are released singly into each glass vial or cell, where they are then cultured for 10 days at 25°C. The fifth day marks the start of the larvae's death. Individual cadavers are collected, transported to disposable 500 ml containers, and immediately frozen until processing. This processing method is comparable to that of *Helicoverpa armigera* Tataksuka *et al.* (2007).

The cadavers are homogenised in sterile, ice-cold water using a liquidiser or a cool, all-glass pestle and mortar at a ratio of 1:2.5 (w/v). The material is repeatedly washed in water after being filtered through double stratified cloth. For this purpose, the water ratio to be used is 1:7.5-12.5 (w/v) for the first weight of the processed corpse. The remaining mat on the cloth is thrown away, and also the filtrate may be semi-purified by differential activity.

To remove debris, the filtrate is centrifuged for 30 to 60 seconds at 500 rpm. Next, the supernatant is centrifuged at 5,000 rpm for 20 minutes. The solid occlusion bodies (POB)-containing pellet is afterwards centrifuged in sterile water three times at a low rate before being subjected to high-speed activity. In order to calculate the strength of the letter box within the refined suspension, the pellet is eventually collected and generated up to a famed volume in water

The *in vivo* production process is comprised of three steps:

1. Infect larvae susceptible species,
2. Grow infected larvae for a while to allow the infection to develop and the virus to replicate
3. Harvest the larvae and extract the virus

However, it has not been discovered to be simple in practise to maintain a prolonged output of virus. Effective production can only be reached and maintained with well-trained personnel, sufficiently planned production procedures, the suitable equipment, and a high standard of process quality control.

Major Insect Pests and Their Management By NPV

Spodoptera litura

Spodoptera litura (Lepidoptera: Noctuidae) is a polyphagous pest of different crops and it has attended global importance due to its damaging range and insecticide resistance problems. SlNPV can be good alternate for the control of *Spodoptera litura* on cabbage.

It is a multi-crop pest that has gained attention due to its destructive range and chemical resistance issues. SlNPV is a viable alternative for *Spodoptera litura* management on cabbage.

Mode of Application of Slnpv

500 lit/ha is the rate at which SlNPV is applied in the field. It is necessary to spray twice; the first time is 45 days after transplanting cabbage in the field, and the second time is 60 days afterwards.

Mode of Application of Hanpv

A concentration of 1.15x107 OBs/ml of HaNPV is sprayed using a knapsack resulted in a100% reduction in *H. armigera* larval infestation within 7 days on tomato plants in a hot house environment. A 10-fold lower concentration, 1.15 x 106 OBs/mL, resulted in a 100%reduction within 16 days.

Worldwide, the American bollworm (ABW), *Helicoverpa armigera* (Hubner), is regarded as a significant pest of cotton, *Gossypium hirsutum*. It evolved resistance against numerous insecticides as a result of its destructive eating behaviour and constant intake of the same poisons. In order to combat this pest, a combination of bio- and synthetic-pesticides must be applied. The entomopathogenic viruses, like nuclear polyhedrosis virus (NPV), a member of baculoviruses could be possible candidates for more effective ABW control.

Nawaz *et al.* (2019) compared the effectiveness of NPV and Spinosad 240SC (at a concentration of 250 mL per hal) against ABW in the controlled environment. Different NPV and Spinosad concentrations were applied to the ABW both singly and in combination with 0.1% Spinosad. According to the findings, mortality rates at 1 109, 1 108, 1 107, 1 106, and 1 105 POBs were 95%, 92%, 84%, 82%, and 78%, respectively, with the highest concentrations showing the greatest mortality (95%) rates. When mixed with food, spinosad causes 100% death at a concentration of 0.8%, followed by 50.87%, 42.10%, 26.31%, and 22.80% mortality at concentrations of 0.4%, 0.2%, 0.1%, and 0.5%, respectively. The findings of this study demonstrated that NPV is a useful strategy for microbial management of ABW.

The *Spodoptera litura* (F.) (Lepidoptera: Noctuidae) Oriental leaf worm moth is one of the most ferocious insect pests in a number of commercially significant crops, including cotton. It now resists the majority of common chemical pesticides. Evaluation of a combined application of bio- and synthetic pesticide against this insect is crucial. For the control of this pest, nuclear polyhedrosis virus (NPV) can be a powerful substitute.

The goal of the current investigation was to assess the effectiveness of NPV and emamectin benzoate (Proclaim® 19EC) in vitro against three geographically diverse populations of S. litura. A combination of emamectin benzoate (EB 0.1 ppm) and three different concentrations of NPV (NPV-1 2 109, NPV-2 3 109, and NPV-3 4 109 POB ml1) were used to treat second and fourth larval instars.

The findings revealed that across all experiments, NPV-3 + EB had the greatest mortality rate (83.28%), followed by NPV-2 + EB, NPV-1 + EB, EB, NPV-3, NPV-2, and NPV-3. The pathogenicity of the used infections was found to be directly connected to reductions in pupation, adult emergence, and egg eclosion. The findings of this study showed that combination application of NPV + emamectin benzoate was an efficient method for biorational control of *S. litura* (Yasin *et al.*, 2020).

Lepidopteran pests infected by nucleopolyhedroviruses (NPV, *Baculoviridae*), which have a proven track record as safe and efficient biological pesticides. Here, we outline a novel strategy for creating NPV-based pesticides. This innovation requires taking advantage of both the distinctive mode of transmission of these viruses as collective infectious units and the genotypic variety found in natural virus populations. The selection of genotypic variants, mixing, coinfection, and intraspecific coocclusion of variants within viral occlusion bodies are all presented as steps in a ten-step process. This method can be utilised to create highly harmful virus formulations for pest management using two instances.

Due to the NPV-based insecticides' limited host range, this technology has recently been modified to produce specially formulated interspecific virus mixtures that can be used to control complexes of lepidopteran pests on specific crops, provided that a common host species is available for virus production. This method of NPV-based pesticide creation has the potential to be used in a variety of NPV-pest pathosystems. Williams *et al.* (2022).

Conclusion

The repeated use of synthetic pesticides against major pests could be the main reason of pest resurgence, and this study results would provide useful insight to build a framework for future investigations and to reduce the use of toxic chemicals for the control of insect pests of major crops. The study methodology for NPVs has developed over the past 25 years towards a more ecologically holistic approach with industry concerns. Although viral pesticides currently only account for a very small fraction of plant protection, their significance was recognised. Even though NPVs have gained popularity as safer alternatives to chemical insecticides, they continue to confront challenges in terms of manufacture, marketing, and use.

Effective multidisciplinary research, governmental, corporate, and human relationships are crucial. Farmers, extension agents, and policymakers must have in-depth knowledge of biopesticides. Ineffective rules may result in poor performance and quality of products as well as a drop in confidence among consumers. There is tremendous potential for NPVs that can function well across a wider range of situations and hosts. Prioritize research to improve bio-agent integration into production systems, such as by alternating them with chemical pesticides and creating efficient bio-models for them.

Referencess

Arora, R. and Padmanaban, N. (2002). Field Evaluation of Native Nucleo-polyhedrovirus for the Management of Sunflower Head borer, Helicoverpa armigera (Hiibner). Annals of Plant Protection Sciences, 10(2): 184-187.

Cunningham, J.C. (1995). Baculoviruses as microbial insecticides. Novel approaches to integrated pest management, 261-292

Deb, S., Chakraborty, S., Hazra, N., SAHA, A. and Chakrabopy, C. (2015). Grasserie in Silkworm, Bombay Mor/L In West Bengal. Journal of Plant Protection and Environment, 12(1): 19-25

Harrap, K.A. (1970). Cell infection by a nuclear polyhedrosis virus. Virology, 42(2): 311-318

Nawaz, A., Ali, H., Sufyan, M., Gogi, M.D., Arif, M.J., Ranjha, M.H. and Ghramh, H.A. (2020). Comparative bio-efficacy of nuclear polyhedrosis virus (NPV) and Spinosad against American bollwormm, Helicoverpa armigera (Hubner). Revista Brasileira de Entomologia, 63: 277-282

Takatsuka, J., Okuno, S., Ishii, T., Nakai, M. and Kunimi, Y. (2007). Productivity and quality of polyhedral occlusion bodies of a nucleopolyhedrovirus harvested from Spodoptera litura (Lepidoptera: Noctuidae) larvae. Applied Entomology and Zoology, 42(1): 21-26

Williams, T., López-Ferber, M. and Caballero, P. (2022). Nucleopolyhedrovirus coocclusion technology: A new concept in the development of biological insecticides. Frontiers in Microbiology, 12: 810026

Yasin, M., Qazi, M.S., Wakil, W. and Qayyum, M.A. (2020). Evaluation of nuclear polyhedrosis Virus (NPV) and emamectin benzoate against Spodoptera litura (F.) (Lepidoptera: Noctuidae). Egyptian Journal of Biological Pest Control, 30(1): 1-6

13

Mass Production Technology of *Trichoderma* spp.

***Hariom Dwivedi*[1*]*, Alka Sharma*[2] *and P.K. Verma*[1]**

[1]*Department of Plant Pathology, Banda University of Agriculture & Technology Banda-210001, Uttar Pradesh, India*

[2]*Department of Biosciences (Microbiology), Integral University, Lucknow-226026 Uttar Pradesh, India*

Introduction

Trichoderma spp. are free living, filamentous fungi that multiply rapidly on many substrates such as soil, decaying wood materials and as root colonizers. *Trichoderma* spp. have a high potential for reproduction and sporulation, as well as competitive ability and saprophytic survival. The escalating need for eco-friendly and sustainable agriculture has driven extensive research on biological control agents like *Trichoderma*. These beneficial fungi exhibit potent antagonistic activity against plant pathogens, offering a promising alternative to conventional chemical pesticides. This review aims to provide insights into the mass production of *Trichoderma* and its utilization as a biocontrol agent for managing plant diseases in agriculture. *Trichoderma* strains in natural habitats may provide more predictable and effective control of plant diseases. For effective biological control of soil borne plant pathogens, a major consideration is antagonist proliferation after introduction into the soil or rhizosphere. Successful antagonist should produce inoculums in excess to survive and proliferate in soil and rhizosphere. *Trichoderma* spp. are biocontrol agents extensively used in management of fungal diseases of crop plants exhibiting mycoparasitism against a wide range of plant pathogens. Their usage has been very much successful against soil borne diseases for which no resistant sources have been identified in the plants. The National Farmer Policy (2007) has strongly recommended the promotion of biopesticides for increasing agricultural production, sustaining the health of farmers and environment. Identification and quantification of ecological stresses affecting the establishment and the population dynamics of introduced *Trichoderma* strains in natural habitats may provide more predictable and effective control of plant diseases. *Trichoderma* spp. (and probably other organisms, such as non-pathogenic *Fusarium* and *Rhizoctonia* strains, mycorrhizae, and other fungi) infect roots and have intrinsic ability to be plant pathogens, but their infection are limited to root superficial cells that is a remarkable phenomenon. It also includes the clause that biopesticides would be treated at par with chemical pesticides in terms of support and promotion. Bioagents like, *T. viride, T. harzianum* and *T. virens* are being successfully used for the control of some of the dreaded diseases like, foot rot of black pepper, root rots of pulses, damping off, collar rots and Fusarium wilts of horticultural crops.

Characterization of *Trichoderma* species

Trichoderma encompasses a diverse group of fungi with substantial variation in their biocontrol efficacy. Their mycoparasitic nature and ability to produce an array of secondary metabolites contribute to their antagonistic activity against phytopathogens. *Trichoderma* strains are characterized based on their morphological, molecular, and biochemical traits, enabling the selection of the most suitable strains for specific biocontrol applications.

To develop a successful *Trichoderma* formulation, *Trichoderma* should possess:

1. High rhizosphere competence.
2. Enhanced plant growth.
3. High competitive saprophytic ability.
4. Ease for mass multiplication.
5. Compatible with other bioagents.
6. Excellent and reliable control.
7. Safe to environment.
8. Broad spectrum of action.
9. Should tolerate desiccation, heat, oxidizing agents and UV radiations.

Environmental Factors

Temperature plays a crucial role in the mass multiplication of *Trichoderma*. The ideal temperature range for *Trichoderma* growth is around 25±2°C, while growth is significantly reduced below 10°C or above 35°C. Other environmental factors such as humidity and pH also influence the growth and viability of *Trichoderma* cultures. Proper control of these factors ensures optimal biomass production and viability of the biocontrol agent.

Methods of mass multiplication of *Trichoderma*

Trichoderma spp. are beneficial fungi used in agriculture and horticulture for their biocontrol and plant growth-promoting properties. Mass multiplication of *Trichoderma* involves propagating large quantities of these fungi for commercial or large-scale applications. Mass production of *Trichoderma* involves various fermentation methods, including liquid fermentation, solid fermentation, and semi-solid fermentation. Liquid fermentation offers efficient and rapid growth of *Trichoderma* in submerged cultures, while solid fermentation relies on surface colonization of solid substrates. Semi-solid fermentation provides a balance between the advantages of liquid and solid fermentation. Additionally, formulation techniques like talc-based and press mud-based formulations improve shelf life and ease of application for commercial products.

Liquid fermentation

Liquid fermentation is conducted in bioreactors under optimized conditions of temperature, pH, and aeration. The liquid medium is enriched with nutrients to support *Trichoderma* growth and biomass production. Liquid formulation involves cultivating *Trichoderma* in liquid media such as nutrient broths or fermentation tanks. The fungi are allowed to grow and proliferate in the liquid medium. Once the desired biomass concentration is reached, the liquid formulation can be used directly or further processed to produce concentrated suspensions or spore solutions. Liquid formulations are convenient for application through irrigation systems or foliar sprays.

Fig: 1. *Trichoderma* grown on liquid media (PDB Broth)

Solid fermentation

In solid fermentation, *Trichoderma* is cultivated on solid substrates such as grains, agricultural residues, or industrial by-products. The fermentation process allows *Trichoderma* to colonize the surface of the substrate. The process involves inoculating the chosen substrate with *Trichoderma* spores or mycelium and allowing them to grow and colonize the substrate. After sufficient growth, the *Trichoderma* biomass is harvested and dried to form solid pellets or granules, which can be used as biocontrol agents or soil amendments.

Sugarcane baggase

The sugarcane baggase is easily or almost freely available substrate for mass multiplication of fungal agent. The sugarcane baggase contains much more amount nutrition for growth of fungal agent *Trichoderma candidum*. The highest CFU count was recorded on solid.

Fig. 2. *Trichoderma* on solid media (PDA Media)

Semi-solid fermentation

Semi-solid fermentation involves the cultivation of *Trichoderma* on substrates with intermediate moisture levels. It offers advantages like higher spore yield and simplified downstream processing.

Talc-based formulation

Talc-based formulations are produced by mixing *Trichoderma* propagules with talc powder, providing an easy-to-apply, dry, and durable product. *Trichoderma* is grown in the liquid medium is mixed with talc powder in the ratio of 1:2 and dried to 8% moisture under shade. The talc formulation of *Trichoderma* has shelf life of 3 to 4 months. It has become quite popular in India for management of several soil-borne diseases of various crops through seed treatment at 4 to 5 g/kg seed. Several private industries produce large quantities of talc formulations in India for supply to the farmers.

Vermiculite-wheat bran-based formulation

This formulation method combines vermiculite and wheat bran as carriers for *Trichoderma* biomass. Vermiculite is a mineral that retains moisture and provides aeration, while wheat bran acts as a nutrient source for the fungi. The *Trichoderma* spores or mycelium are mixed with the carrier materials, and the formulation is then processed into granules or pellets.

Press mud-based formulation

Press mud, a by-product of the sugar industry, can serve as a carrier for *Trichoderma* propagules. This cost-effective formulation enhances nutrient content and promotes sustained release.

Village level mass multiplication of *Trichoderma*

Trichoderma can be easily produced in rural areas by the farmers themselves using simple and cheap materials locally available in the village. *Trichoderma* can be produced three types:

1. Cheap grains (Sorghum, Maize, wheat)
2. Liquid media (Potato jiggery medium)
3. Agricultural by-products (Rice bran, Wheat bran)

Trichoderma grown on sorghum or other grains is powdered and used like talc formulation for disease control. *Trichoderma* grown in liquid media will be used to make the talc formulation of *Trichoderma,* which can be used for, seed treatment, root dip and soil application for disease control.

Uses as plant disease management

Trichoderma-based biocontrol agents have demonstrated efficacy in managing various plant diseases caused by fungi, bacteria, and nematodes. They stimulate plant growth, induce systemic resistance, and directly antagonize pathogens, leading to enhanced crop health and reduced disease incidence. *Trichoderma spp.* has gained significant attention and use in plant disease management due to their biocontrol capabilities. These beneficial fungi can effectively suppress plant pathogens and contribute to sustainable agriculture. Here are some of the key ways *Trichoderma* is used for plant disease management:

1. Biological Control of Soil-Borne Pathogens

Trichoderma species are well-known for their antagonistic activity against various soil-borne pathogens, such as *Fusarium, Rhizoctonia, Pythium,* and

Sclerotinia. When applied to the soil, *Trichoderma* colonizes the rhizosphere (root zone) and competes with the pathogenic fungi for space and nutrients. It also produces antifungal compounds that inhibit the growth of pathogens, thus reducing disease incidence.

2. Seed Treatment

Trichoderma-based formulations can be used for seed treatment. Coating seeds with *Trichoderma* spores or mycelium helps establish a beneficial microbial population around the seedlings' roots from the beginning. This early protection can prevent soil-borne pathogens from infecting the emerging seedlings.

3. Root Drench and Soil Drench

Trichoderma products in liquid formulations can be applied as root drench or soil drench to protect plants from soil-borne diseases. When applied to the base of plants or directly to the root zone, *Trichoderma* establishes a protective barrier and prevents pathogens from infecting the root system.

4. Foliar Spray

Some *Trichoderma* strains have the ability to colonize above-ground plant parts. Foliar sprays of *Trichoderma*a-based products can enhance plant defenses against foliar diseases caused by certain pathogens. The presence of *Trichoderma* on the leaf surface can induce systemic resistance, priming the plant to respond more effectively to pathogen attacks.

5. Post-Harvest Disease Management

Trichoderma-based products can also be used for post-harvest disease management. Applying these formulations on harvested fruits or vegetables can suppress the growth of post-harvest pathogens, extending the shelf life of produce and reducing post-harvest losses.

6. Inducing Systemic Resistance

Trichoderma spp. can trigger the plant's systemic resistance mechanisms. When a plant is exposed to Trichoderma, it initiates a signaling cascade that leads to the activation of defence genes, making the plant more resistant to various diseases.

7. Nematode Management

Certain *Trichoderma* strains have nematode-suppressive activity. They can help control plant-parasitic nematodes, either directly through predation or indirectly by stimulating the plant's defence against nematode infection.

Commercial product for biocontrol agent

The successful mass production of *Trichoderma* has facilitated the development of commercial biocontrol products. These formulations offer convenient and practical solutions for integrated disease management practices in modern agriculture.

***Trichoderma harzianum strain T22*:** This strain of *Trichoderma* is commonly used as a biofungicide to protect against various plant pathogenic fungi.

***Trichoderma asperellum*:** Another popular strain used in commercial biofungicides to control plant diseases.

***Trichoderma virens*:** This strain is known for its ability to colonize plant roots and protect against soil-borne pathogens.

***Trichoderma koningii*:** Used in biofungicides to target diseases caused by soil-borne pathogens.

***Trichoderma viride*:** A widely used strain in commercial products for its effectiveness against plant pathogens.

Conclusion

Trichoderma species possess immense potential as biocontrol agents for managing plant diseases in agriculture. Their mass production through various fermentation methods and formulation techniques has opened up new avenues for sustainable crop protection. Emphasizing the commercialization and adoption of *Trichoderma*-based products can contribute to environmentally friendly and economically viable strategies for plant disease management. *Trichoderma* spp. possess many qualities and they have great potential use in agriculture such as amend abiotic stresses, improving physiological response to stresses,alleviating uptake of nutrients in plants, enhancing nitrogen-use efficiency in different crops, and assisting to improve photosynthetic efficiency. Nonetheless, continuous efforts are necessary to raise awareness, educate farmers, and integrate *Trichoderma* biocontrol into mainstream agricultural practices to maximize its potential for disease management and overall crop health. Embracing *Trichoderma* as a biocontrol agent holds the promise of achieving sustainable and resilient agricultural systems while preserving the environment for future generations.

References

Baker K.F. and Cook R.J. (1974). Biological control of plant pathogens, Wh Freeman and co, San Francisco, American Phytopathological Society. St. Paul. MN. 433 pp

Grondona I., Hermosa R.M., Tejada M.D., Gomis P.F., Mateos P.D., Bridge E. and Garcia-acha I. (1997). Physiological and biochemical characterization of Trichoderma harzianum a biological control agent against soil borne fungal plant pathogens. App Envir Microbiol 63(8): 3189

Harman, G.E., Howell, C.R., Viterbo, A., Chet, I. and Lorito, M. (2004). Trichoderma species opportunistic, avirulent plant symbionts. Nat Review Microbiol. 2(1): 43-56

Howell, C.R. (2013). Mechanisms employed by Trichoderma species in the biological control of plant diseases: The history and evolution of current concepts. Plant Diseases 87(1): 5-9

Jeyarajan R. and Nakkeeran S. (2000). Exploitation of microorganisms and viruses as biocontrol agents for crop disease mangement. In: Biocontrol Potential and their Exploitation in Sustainable agriculture, (Ed. Upadhyay et al.,) Kluwer Academic/ Plenum Publishers, USA. pp. 95-116

Kumar, S., Thakur, M. and Rani, A. (2014). Trichoderma: Mass production, formulation, quality control, delivery and its scope in commercialization in India for the management of plant diseases. African Journal of Agricultural Research, 9(53): 3838-3852

Ramanujam, B., Prasad, R.D., Sriram, S., and Rangeswaran, R. (2010). Mass production, formulation, quality control and delivery of Trichoderma for plant disease management. The Journal of Plant Protection Sciences, 2(2): 1-8

Sinha, B., Rajendran, P. and Devi, P.S. (2022). Mass Production of Trichoderma from Agricultural Waste and Its Application for Plant Disease Management. In Handbook of Solid Waste Management: Sustainability through Circular Economy (pp. 619-633). Singapore: Springer Nature Singapore

Trina, S., Tampakleima, A., Rajlakshmi, P. and Sinha, M. (2016). Biocontrol potential of Trichoderma spp. against phytopathogens. In Plant Health Under Biotic Stress (pp. 37-65). Springer, New Delhi

Waghunde, R.R., Shelake, R.M., and Sabalpara, A.N. (2016). Trichoderma: A significant fungus for agriculture and environment. African Journal of Agricultural Research, 11(22): 1952-1965

14

Role of Semiochemicals in Insect Management

Suriya S.[1*], Sadhana V.[2], Santhoshini E.[3] and Sekar S.

[1]*Division of Entomology, Sher-e-Kashmir University of Agricultural Sciences and Technology, Srinagar, Jammu and Kashmir-190025*

[2]*Department of Entomology, Tamil Nadu Agricultural University Coimbatore-641003 Tamil Nadu*

[3]*Department of Plant Pathology, University of Agricultural Sciences, Dharwad Karnataka-580005*

Introduction

Semiochemicals, which can refer to both intraspecies and interspecies communication chemicals, was first coined by Law and Regnier in 1971 (simeon-marker or signal). Semiochemicals are chemical signals that cause different behavioral reactions in organisms. In most groups of animals, chemical communication seems to be the main means of information transfer. Since 1971, the term "semiochemical" has been in use. It is derived from the Greek word "semeon," which meaning "sign" or "signal. Semiochemicals are communication agents that alter the behaviour of living things by carrying information between them (Dicke M and Sabelis M W, 1988). They are emanated by one individual and result in a reaction in another. According to Krieger and Breer (1999), the majority of invertebrates use smell as their primary sensory modality to perceive their surrounding environment. Pheromones and allelochemicals are two categories of semiochemicals that function inside and between species, respectively. Allelochemicals can be further classified into signals that are beneficial to the emitter (allomones), the receiver (kairomones), or both (synomones). Pheromones have evolved for communication reasons and can be classified as sex, alarm, aggregation, or territory marking signals. (Nordlund *et al.,* 1981). Chemical communication is often the main method by which animals exchange information, as observed in various animal groups. The presence of a highly developed sense of smell allows for the detection of even faint signals from prospective partners, as well as distinguishing between favourable and unfavourable environments for feeding, mating, and egg-laying. Semiochemicals refer to organic substances utilized by insects to communicate particular chemical messages that impact behaviour or physiology (El-Sayed, 2015). They play a crucial role in modifying the actions and physical processes of these insects.According to Bruce *et al.* (2015), the attraction of insects to plants and other hosts depends on the identification of particular Semiochemicals or particular combinations of Semiochemicals.

Historical background

Numerous animals were used in the pheromone research. In fact, more than 80% of all species surveyed have been the subject of studies on insect semiochemicals, particularly pheromones. This is likely because there is growing interest in their possible application in the control of insect pests.

The French naturalist Jean Henri Fabre wrote the earliest works on the possible function of pheromones when he noted in 1870 that the female peacock moth, *Saturniapyri*, could draw dozens of male moths from great distances (Schwarcz, 2003). In 1870, Joseph Linter observed that female spice bush silk moths attracted 50 males, and he recommended that chemists step up to assist entomologists in the identification of pheromones.

I) Allelochemical

A substance that is significant to organisms of a species different from source, for reasons other than food.

Allomone: A chemical generated by an organism that, when it comes into contact with a member of a different species, causes the recipient to exhibit behavioural or physiological responses that are advantageous to the emitter but unfavourable to the recipient in terms of adaptation.

Kairomone: A chemical that an organism produces or acquires that, when it comes into contact with a member of another species, causes the recipient to react in a way that is advantageous for receiver's ability to adapt.

Synomone: A substance produced or obtained by an organism that, when it comes into touch with a member of a different species, causes the recipient to exhibit behaviour or experience physiological changes that are advantageous to both the emitter and the receiver in terms of adaptation.

Apneumone: Achemical released by non-living substance that is beneficial to the receiver but detrimental to other organism in the substance.

II) Pheromones

A chemical that an organism secretes outside the body and that elicits a particular response in a receiving organism of the same species.

a) Sex pheromones

Sex pheromones are only released by one sex and cause the other sex to act in ways that promote mating. The majority of the time, females release them, while males can also release them. More than 150 kinds of insects have been reported to release sex pheromones, and roughly 50 species are produced by male insects. When the two sexes of an insect are brought together, aphrodisiacs are compounds that help with courtship. Aphrodisiacs are frequently made by male.

Differences between male and female produced pheromones

S.No	Properties	Female sex pheromone	Male sex pheromone
1	Range	Acts at a long range. Attracts males from longdistance	Acts at a short distance
2	Role of other stimuli	Playlessrole	Visual and auditory stimuli play majorrole
3	Action elicited in the other sex	Attracts and excites males to copulate	Lowers females resistance to mating
4	Importance in IPM	Moreimportant	Lessimportant
5	Example	Brinjal shoot and fruit borer	Cotton boll weevil

Chemical structure of sex pheromones of some insect pests

S. No	Insect name	Scientific name	Chemical structure
1	Red gram pod borer	*Helicoverpa armigera* (Hubner)	(Z)-11-hexadecacenal and (Z)-9-hexadecenal (9:1)
2	Pink bollworm	*Pectionophora gossypiella* (Saunders)	(Z, Z)- 7,11hexadecadienyl acetate and (Z, E)-7,11 hexadecadienyl acetate (61:39)
3	Tobacco cutworm	*Spodoptera litura* (Fabricius)	(Z, E)-9,11- tetradecadienyl acetate and (Z, E)-9,12 tetradecadienyl acetate (9:1)
4	Yellow stem borer	*Scirpophaga incertulus* (Walker)	(Z)-9-hexadecacenal and (Z)-11-hexadecenal (1:3)
5	Pink stem borer	*Sesamia inferenes* (Walker)	(Z)-11-hexadecacenal acetate and (Z)-11- hexadecen-1-ol and (Z)-11-hexadecanal (4:1:0.1)
6	Armyworm	*Mythimna separata* (Walker)	(Z)-11-hexadecacenal acetate and (Z)-11- hexadecacenal (5:100)
7	Cutworm	*Agrotis ipsilon* (Hufnagel)	(Z)-7- dodecenylacetate, (Z)-9-tetradecenyl acetate and (z)-11-hexadecenyl acetate (1.79:6.12:3.66)

8	Angoumois grain moth	*Sitotroga cerealella* (Olivier)	(Z, E)- 7,11hexadecadienyl acetate and (Z, E)-7,11 hexad
9	Indian meal moth	*Plodia interpunctella* (Hubner)	(Z, E)-9,12 tetradecadiene 1-ol- acetate and (Z, E)- 9,12 –tetradecadiene 1-ol(3.2:1.2)
10	Codling moth	*Cydia pomonella* (Linnaeus)	(E, E)-8,10-dodecadien-1- ol
11	Gypsy moth	*Lymantria dispar* (Linnaeus)	(7R,8S)-cis-7,8-epoxy-2-methyloctadecane and (7R,8S)-cis-7,8 –epoxy2-methlocatadec-17-ene
12	Cabbage looper	*Trichoplusia ni* (Hubner)	(Z)-7- dodecenyl acetate
13	Tomato pin worm	*Keiferialycopersicella* (Walsingham)	(E)-4-tridecenyl acetate
14	Brinjal fruit and shoot borer	*Leucinodes orbonalis* (Guenee)	(E)-11-hexadecenyl acetate, (E)-11-hexadecen-1-ol
15	Red palm weevil	*Rhynchophorous ferrugineus* (Olivier)	(4S,5S)-4-methyl -5- nonanol and 4 – methyl-5- nonanone

b) Aggregation pheromones

These pheromones are particularly prevalent in some species of beetles and they drive insects to gather or congregate at food sources, reproductive habits, and hibernation places. They are particularly well studied in relation to bark beetles like *Dendroctonus* spp. that attack trees. Both men and women find them attractive. They typically function over a long period of time and have the ability to draw thousands of people. Aggregation pheromones have chemical structures that are more complicated, less stable, and less easy to manufacture and application. They also elicit considerably more complex behaviour that is less responsive to manipulation.

c) Trial marking pheromone

Trail pheromones are produced by ants, termites, and some lepidopteran insect larvae that are foraging. They mostly serve as a guide for other colonists regarding where to obtain necessities. In order to create appealing foraging areas and to scent-mark unproductive food sources, bees use trail pheromones during foraging. As a rule, trail pheromones are less volatile than alarm pheromones. It may be possible to use bumble bees for pollination more effectively by identifying and synthesising their trail pheromone.

d) Alarm pheromones

These are typical of social insects that typically congregate, including ants, bees, and aphids. This kind of pheromone has three main purposes: to alert conspecifics, to elicit a defensive reaction, and to spur avoidance. For instance, a honey bees stinger that was left inside the victim's body generates an alarm scent that draws other bees and prompts them to sting.

e) Host-marking pheromones

By preventing pests from landing, eating, or ovipositing on their host plants, these pheromones lessen intraspecific competition. Being one of the few pheromones that repel insects rather than attracting them, they are so rare. The Coleopteran, Lepidoptera, Diptera, Homoptera, Orthoptera, and Hymenoptera are only a few of the insect orders from which they are known. The females of cherry fruit flies, *R. cerasi* (Linnaeus), and apple maggot, *Rhagoletis pomonella* (walsh), which lay their eggs in fruit, mark the surface to ward off other females. Numerous insects, including tsetse flies, house flies, and other dipteran species, produce mating-deterrent pheromones.

f) Recruitment pheromones

These are often utilised by social insects to keep the group in order, recruit new members, and plan their actions. In honeybees, worker bees emit the vasonov pheromone to attract new workers from outside the hive and to direct returning forager bees back to the colony.Bees elevate their vasonov gland-containing abdomens and vigorously fan their wings in order to produce this hormone. When bark beetles locate a good tree for colonisation, we can also discover these pheromones in them.

g) Recognition pheromones

To differentiate colony members from non-colony members, social insects such as ants, bees, and termites are used. These pheromones often consist of a mixture of straight or branched-chain simple hydrocarbons. One of the most significant pheromones discovered is the termite egg recognition pheromone (TERP), which is known to powerfully trigger the worker conduct of egg-carrying and grooming.

h) Retinue pheromones

These compounds, which the honey bee queen is said to release, are supposed to trigger the retinue attraction that motivates workers to feed, groom, and collect the queen's pheromones and spread them throughout the colony. The queen retinue pheromone contains nine distinct components, according to research.

Semiochemicals and their applications in insect pest control

Sex pheromones, primarily those of lepidopteran insects, which were among the first to be recognised and synthesised, are currently the semiochemicals that are utilised most frequently in the management of insect pests. Pheromones offer important properties such approximate species specificity, potency, and low toxicity when it comes to pest management. The semiochemicals serve not as 'killing agents' per se but rather as monitoring instruments (to optimise the timing of conventional chemical controls), therefore they are not a straight replacement for chemical insecticides. According to Reddy *et al.* (2005), the objectives of employing semiochemicals in pest management include

- Tracking pest population trends
- To ascertain if control is allowed
- To change the behaviours of an insect pest or its foes to the pest's harm.

Advantages of using semiochemicals are

- They only negatively impact the targeted pest.
- They are comparatively non-toxic and only needed in small quantities.
- Environmentally safe and non-persistent.

According to Soroker *et al.* (2015), Semiochemicals have been utilised to manage insect pests for more than a century. Semiochemicals called insect sex pheromones are frequently utilised to control insect pests, notably those from the order Lepidoptera. It is also common practise to use aggregation pheromones from the Coleoptera order to manage economically significant agricultural insect pests. The carob moth *Ectomyelois ceratoniae*, the autumn armyworm *Spodoptera frugiperda*, the tomato leaf miner *Tutaabsoluta*, fruit flies *Bactrocera* sp., mountain pine beetle (MPB) *Dendroctonus ponderosae*, Asian citrus psyllid *Diaphorina citri*, and the red palm weevil (RPW) have all been successfully controlled by semiochemicals. Due to their natural origin, low environmental persistence, and species specificity—all of which greatly increase their toxicity to non-target organisms-semiochemicals are regarded as safe and environmentally benign substances.

Semiochemical-based pest control techniques Attract and kill (A&K)

The approach, also known as attract and kill, attract and infect, or attract and sterilise, uses an attractant or semiochemical to entice an insect to a point source holding a killing agent (insecticide, pathogen, or sterilant), as the name implies. By killing the target insect, lowering its fitness and fecundity, or rendering it inoperable by infecting it with a disease, the approach lowers the insect population.

According to Wizgall *et al.* (2010), semiochemicals have a practical use in detecting the presence and the amount of pest populations. The monitoring system assists in making decisions about pest control strategies that keep the pest population below the point at which it becomes economically unviable. Kairomones or pheromone baits or traps can be used for this. The biological control agent *Rhizophagus grandis*, a predator of the spruce bark beetle *Dendroctonus micans* population, has been observed using kairomone-based tarps often (Hosking *et al.*, 2003). Pheromone bait traps are more successful in monitoring insects than kairomones because pheromones are strong attractants and species-specific attractants (El-Sayed *et al.*, 2006). Insect pest damage can be evaluated using a pheromone-based monitoring system, together with population trends and abundance. The size of trap captures is used to establish catch thresholds, which are then used to time the application of control measures or to make that decision (Meurisse *et al.*, 2008).

Monitoring

The monitoring system aids in the choice of control measures to limit the pest population below the point at which it becomes unprofitable. Insect behaviour is now monitored by sex pheromones, aggregation pheromones, and kairomones. As a predator of the *Dendroctonus micans* population of the spruce bark beetle, *Rhizophagus grandis,* a biological control agent, kairomone-based traps have been most frequently employed to monitor this population (Hosking *et al.*, 2003). In addition to monitoring the presence of invasive species and preventing their establishment and spread, these pheromone traps are utilised to identify low-density populations (Trematerra, 2012).

Mating Disruption Strategy

In order to disrupt mating, a large quantity of artificial pheromones is released into the pest's habitat in a quantity that makes it difficult for the male to find the female, decreasing the organism's likelihood of reproduction and, as a result, lowering the prevalence of the insect in the following generation. This method was initially proven for the cabbage looper, but it is now a widely acknowledged choice for controlling a variety of lepidopteran pests of fruits and vegetables. After undertaking an extensive disruption programme, hot spots and edge infestations of the Oriental fruit fly moth were eliminated in Australia.According to some reports, the gipsy moth (*Lymantria dispar*) in North American forests, the codling moth (*Cydia pomonella*) in apple and pear trees worldwide, and the grapevine moth (*Lobesiabotrana*) in the EU and Chile were the three species with the highest land area under mating disruption technique (Witzgall *et al.*, 2010). In a recent study, the effectiveness of pheromone-based mating disruption strategies of the swede midge (*Contarinia nasturtii*), a member of the dipteran family, was evaluated in small-scale and large-scale fields of broccoli and cauliflower. In the large-scale studies, the crop damage was reduced on average by 91% and by 59% in the case of broccoli (Samitez *et al.*, 2012).

Mass Trapping Strategy

Mass trapping tries to capture a sizable fraction of a pest population prior to mating, oviposition, or feeding in order to stop crop damage. Using this method successfully requires a highly effective trap and an extremely alluring lure. The lure should be extremely active, eventually outcompeting the attracted substance that occurs naturally. Lepidoptera require that males be captured prior to mating, and this is most likely to be successful with insects that only pair once. *Leucinodes orbonalis*, a moth that damages the fruit and shoots of brinjal plants, has been controlled as a result of a combination of pest population decline and increased influence of natural enemies, whose numbers increased after insecticide usage was stopped (Tamiru and Khan 2017). When it comes to coleopterans, aggregation pheromone-based trapping tries to lower the population of both sexes before eggs are deposited or harm is caused by adults being fed. Recently, the mass trapping of the New Guinea sugarcane weevil, *Rhabdoscleus obscurus*, with ethyl acetate, cut sugarcane, and insecticide (permethrin) was demonstrated in Guam (Reddy *et al.*, 2005), as well as with the boll weevil, *Anthonomus grandis*, and the banana weevil, Cosmopolites sordidus (Miller *et al.*, 2010).

Conclusion

In the last two to three decades, the use of semiochemicals in pest management has advanced quickly due to a number of special advantages. These include the fact that they are species-specific and have no harmful effects on organisms other than the target species. Finally, because they are used sparingly while killing a large number of insects, they are undoubtedly cost-effective. This strategy saves manpower because a huge number of insect pests can be brought in from afar and exterminated directly at the door. Semiochemicals can't produce quick effects, hence they can't be used for immediate control measures. Only one sex can be attracted by sex pheromones; the other sex may still be present and still cause harm. Additionally, it is yet uncertain what semiochemicals are for many kinds of insect pests.The widespread usage of semiochemicals, such as tomato pinworm and pink bollworm sex pheromones, increases the risk of resistance forming. It is challenging to create long-lasting controlled release dispensers for chemicals that are frequently very volatile and prone to deterioration.The size of the prospective market, the cost of registration, and the competitiveness of the product's price all have a significant impact on the commercialization of goods based on semiochemicals. The development of goods based on Semiochemicals has undoubtedly been impeded by the big agrochemical companies' lack of economic interest.

References

Bruce, T.J.A.,Wadhams, L.J. and Woodcock, C.M. (2005). Insect host location: a volatile situation. Trends Plant Sci.10(6): 269-274

Dicke, M. and Sabelis M.W. (1988). Info chemical terminology: based on cost benefit analysis rather than origin of compounds. Funct Ecology 2: 131-139

El-Sayed, A.M. (2015). The pherobase: database of insect pheromones and semiochemicals

Krieger and Breer (1999). Olfactory reception in invertebrates. Science 286: 720-723

Miller, J.R., McGhee, P.S., Siegert, P.Y., Adams, C.G. and Huang, J. (2010). General principles of attraction and competitive attraction as revealed by large-cage studies of moths responding to sex pheromone. ProcNatlAcadSci USA 107(1): 22-27

Nordlund, D.A., Jones, R.L. and Lewis, W.J., (1981). Semiochemicals. Their role in pest control. John Wiley and Sons, New York

Reddy, G.V.P., Cruz, Z.T.,Bamba, J. and Muniappan, R. (2005). Development of a semiochemical-based trapping method for the New Guinea sugarcane weevil, Rhabdoscelus obscures. Journal of Applied Entomology. 129: 65-69

Samietz, J., Baur, R. and Hillbur, Y. (2012). Potential of synthetic sex pheromone blend for mating disruption of the swede midge, Contarinianasturtii. Journal of Chemical Ecology. 38: 1171-1177

Schwarcz, J. (2003). Why Moths Are Confused Bachelors? p. 6 The Gazette December 14

Soroker, V., Harari, A. and Faleiro, J.R. (2015). The role of semiochemicals in date pest management. In Sustainable pest management in date palm: Current status and emerging challenges Springer, Cham, 315-346

Tamiru, A. and Khan, Z.R. (2017) Volatile semiochemical mediated plant defence in cereal: A novel strategy for crop protection. Agronomy 7: 58

Witzgall, P., Kirsch, P., and Cork, A. (2010). Sex pheromones and their impact on pest management. Journal of Chemical Ecology. 36(1): 80-100

15

Plant Quarantine and Its Importance in Agriculture

Saurabh Singh[1*], Shailendra Kumar Mishra[2], Himanshu Kumar Gupta[3] and Vivek Singh[4]

[1,2]Department of Entomology College of Agriculture, Banda University of Agriculture and Technology, Banda-210001, Uttar Pradesh, India

[3]Department of Plant Pathology, Veer Bahadur Singh Purvanchal University, Jaunpur-222003, Uttar Pradesh, India

[4]Department of Plant Pathology College of Agriculture, Banda University of Agriculture and Technology, Banda-210001, Uttar Pradesh, India

Introduction

The challenge for all disciplines of agriculture is to increases production and improves quality of agricultural produce. This is applicable to the discipline of plant protection and manage risk. In the past many pathogens are responsible for food scarcities including famines. In addition to endemic problems there are many crop pests which are entered India from other foreign countries because in India did not have an effective control measure (plant quarantine) system to stop the introduction of exotic pests, diseases and weeds. Cottony cushion scale, woolly aphid, San Jose scale, golden cyst nematode of potatoes are some exotic pests introduced into our country and cause extensive damage. In recent yearsthe increases in quantum of import and export of plant commodities is a more possibility of moving insect pests and diseases from their native habitation to new location. The methods of crop protection isto excludes the pests from entering in to new area. The method of exclusion of the pests is enforced by legal measures commonly known as Quarantine. It can be defined "as a legal restriction on the movement of Agricultural commodities for exclusion, prevention or delay in the spread of plant pests and diseases in new location.

Objectives of Plant Quarantine

- Tracing, identification and control of quarantine pests.
- Pest risk analysis, evaluation and determination of quarantine conditions for imported and transit plant shipments.

- Inspection, sampling, testing and undertaking quarantine regulations on imported and transit agricultural shipments.
- Supervision and undertaking post-entry quarantine regulations on plants which are subject to the regulations.
- Inspection, sampling, testing and issuing phytosanitary certificate for exporting plant consignments.

Importance

- Prevent the introduction and spread of non-native pests that are destroy the crops by regulating/restricting the import of materials.
- Provide the safe global trade in agriculture by reliable phytosanitary certificate system to meet exporter of trading partners.
- Plant quarantine is major role to prevent the introduction of foreign pests and diseases into a country, or to intercept and eradicate them before the widespread and well-established.

History of Plant Quarantine in India

1906 - Indian Government in, ordered compulsory fumigation of imported cotton bales to prevent the introduction of the dreaded Mexican cotton boll weevil (*Antonymous grandis*).

1914 - Comprehensive Plant Quarantine, Destructive Insects and Pests Act, (DIP Act)

1946 - Directorate of Plant Protection, Quarantine and Storage, under the ministry of Food and Agriculture were set up. Plant quarantine activity stared with the initiation of plant introduction scheme in the Botany Division at Indian Agricultural Research Institute (IARI) New Delhi.

1949 - Directorate started its quarantine activities at Bombay seaport.

1951 - The first plant Quarantine and Fumigation station in India was formally inaugurated.

1976 - The National Bureau of Plant Genetic Resource (NBPGR) was created.

1978 - The Division of Plant Quarantine was created with Entomology, Plant Pathology and Nematology sections.

1988 - PFS (The Plants, Fruits and seeds order, popularly known as order came into force in 1989.

Milestones of Plant Quarantine

- Destructive Insects and Pests Act 1914
- Locust Warning Organization 1939
- Directorate of Plant Protection, Quarantine and Storage 1946

- National Institute of Plant Health Management 1966
- Insecticides Act 1968
- Implementation of Insecticides Act 1971
- Plants, Fruits, Seeds (Regulation of Import into India) Order 1984.
- Integrated Pest Management (Central IPM Centre's) 1992
- Plant Quarantine Order 2003
- Online Registration of Pesticides 2010
- Online PQ Services (PQIS) 2011
- Pesticides Management Bill 2020
- Plant Quarantine Management System (PQMS) 18 April 2022.
- Comprehensive Registration of Pesticides (CROP) 18 April 2022.
- Principles of plant quarantine as related to international trade

General Principles

1. **Sovereignty** With the aim of preventing the introduction of quarantine pests into their territories, it is recognized that countries may exercise the sovereign right to utilize phytosanitary measures to regulate the entry of plants material capable of harbouring plant pests.
2. **Necessity** Countries shall institute and plant protection centresare restrictive measures by phytosanitary considerations, to prevent the introduction of quarantine pests.
3. **Minimal impact** Phytosanitary measures is consistent with the pest risk involved, and shall represent the least restrictive measures available which result in the minimum impediment to the international movement of people, commodities and conveyances.
4. **Modification** As conditions change or limitations, and as new facts become available, phytosanitary measures shall be modified promptly, either by inclusion of prohibitions, restrictions or requirements necessary for their success, or by removal of those found to be unnecessary.
5. **Transparency** Countries shall publish and disseminate phytosanitary prohibitions, restrictions and requirements and do not hide any wrong information, on request, make available the rationale for such measures.
6. **Harmonization** Phytosanitary measures shall be based on IPPCwhich are possible, on international standards, guidelines and recommendations, developed within the framework.
7. **Equivalence** Countries shall recognize as being equivalent those phytosanitary measures that are not identical but which have the same effect.

8. **Dispute settlement** It is preferable the dispute between two countries regarding phytosanitary measures be resolved at a technical bilateral level (two countries). If such a solution cannot be achieved within a reasonable period of time, further action may be undertaken by means of a multilateral settlement system (more than two countries).

Specific Principles

9. **Cooperation** Countries shall cooperate to prevent the spread and introduction of quarantine pests, and to promote measures for their official control. Cooperation by both importer and exportercountries.
10. **Technical authority** Countries shall provide an official Plant Protection Organization and Institute.
11. **Risk analysis**within the guideline of the IPPC,countries shall use pest risk analysis methods based on biological and economic evidence and, wherever possible, follow procedures developed.
12. **Managed risk** Because some risk of the introduction of a quarantine pest always exists, countries shall agree to a policy of risk management when formulating phytosanitary measures.
13. **Pest free areas** Countries shall recognize the status of areas in which a specific pest does not occur (nonnative pest). On request, the countries in whose territories the pest free areas lie shall demonstrate this status based, where available, on procedures developed within the framework of the IPPC.
14. **Emergency action** Countries may, in the face of a new and/or unexpected phytosanitary situation, take immediate emergency measures on the basis of a preliminary pest risk analysis. Such emergency measures shall be temporary in their application, and their validity will be subjected to a detailed pest risk analysis as soon as possible.
15. **Notification of non-compliance** Importing countries shall promptly inform exporting countries of any non-compliance with phytosanitary prohibitions, restrictions or requirements before the imports the commodities.
16. **Non-discrimination** Phytosanitary measures shall be applied without discrimination between countries of the same phytosanitary status, if such countries can demonstrate that they apply identical or equivalent phytosanitary measures in pest management. Which are benefited for both the countries.

Guidelines for Import of Germplasm

- The plant product or plant material are import from a country where the pathogens is or minimum absent.

- The plant product import from a country with a standard plant quarantine service, so that regularly inspection and treatment is done.
- Planting material are obtained from the safest known source or selected country.
- Planting material obtain without chemical uses seeds so that detection of pathogens is facilitated.
- Import the good, clean, healthy-looking type seeds.
- Obtain an official certificate ofimporting country to free from pests and diseases from the exporting country.
- Smallest possible amount of planting materialis imported. The smaller amount less the chance of it carrying infection. It will also simplify post entry inspection.
- Inspect material carefully on arrival and treat. Don't taken any damage to the plant product.
- If other precautions are not mention, so subject the material to intermediate or post entry quarantine.
- Salvage infected seeds.

Inspection Procedures followed in quarantine station

Inspection method

- **Visual inspection** - To detect sclerotia, nematode galls, buntgalls, smuts, insect infestations, weed seeds insect eggs, inert mater *etc.* by visually seen.
- **X-ray test** - Insect infestation (hidden), e.g., *Callosobruchus spp.*
- **Washing test** - Spores of fungus eggs of insect adhering to seeds, nematode galls.
- **Sedimentation test** - Stem eelworm (*Ditylenchus dipsaci*) (Baerman Funnel Test),
- **Incubation test** - Seed borne fungi/ bacteria (Blotter/agar test).
- **Grow out test** - Seed borne bacteria/viruses/downy mildews.
- **Electron microscopy** - Plant viruses.
- **Serological methods:**
  - a) ELISA (Enzyme Linked Immuno sorbant Assay)
  - b) DIBA (Dot Immuno-binding Assay)
  - c) ISEM (Immuno sorbant Electron Micro Scopy)
  - d) Latex agglutination test.

- **Nucleic acid hybridization,**
- **Polymarised chain reaction (PCR):** These methods are used for molecular testing.

Types of Plant Quarantine

International quarantine International Standards for Phytosanitary Measures are governed by the Secretariat of the International Plant Protection Convention (IPPC) as part of the United Nations Food and Agriculture Organization's (FAO) global programme of policy and technical assistance these standards, guidelines and recommendations to achieve international harmonization of phytosanitary measures, with the aim to facilitate trade and avoid the use of unjustifiable measures as barriers to tradein plant quarantine.

Domestic quarantine Plant quarantine are governed by Destructive Insects and Pests Act (DIP Act) 1914. Therules and regulations issued prohibiting the movement of insects and disease their host from one state to another in India and therules and regulations framed from time to time there under by the Govt. of India (Anonymous, 1975). India brought forward a comprehensive Plants, Fruits and Seeds Order, 1984' which came into force in June 1985 called domestic quarantine (Anonymous, 1985).

Plant quarantine measures: The measures suggested for effective plant quarantine are as follows.

Import control

i) **Embargoes:** Total prohibition of import of certain kind of infected material from particular region /country. This is most effective measure to exclude infected plant material. These are highly infective area which are totally banned by IIPC.

ii) **Inspection of seed lots:** The examination of seed samples must be by the most sensitive and reliable method for detection of dangerous pathogens. A sample may subject to more than one method which are detect by various technique and method.

iii) **Post Entry Quarantine (PEQ):** because it is difficult to detect all types of seed borne pathogens by simple tests, it may be necessary to subject seeds to PEQ. When pathogen are not which are not provide facilitate the environment condition so pathogen are not born in plant materials so require the PEQ. Seeds are subjected to a period of grow in quarantine station under strict supervision in importing countries.

Export control

i) **Field inspection of field crops:** The seed crop is inspected regularly for diseases infected plants are rouged and other weeding practises. The crop should meet requirements of the importing country.

ii) **Inspection of seed lots:** The seed lot is thoroughly examined for the presence or absence of the microorganisms and well cleared before export. The sample should meet the importing country.

iii) **Seed treatment:** The seed lot should be treated by any chemical as per the requirements of the exporting country. However, the treatment should confine to regulations of the importing country.

iv) **Phytosanitary certificate:** Phytosanitary certificates are issued by the exporting country for the healthy planting material are transport along with the seed as per the international plant protection convention.

Phytosanitary Certificate

It is an official document required when transport regulated articles such as plants, plant products or other regulated materials. It is generated from the exporting country's Department of agriculture, horticulture, food or water resources. It can only be issued by an authorized officer from a government department that is authorized by a National Plant Protection Agency (NPPO) and national plant protection station. The agency is in force to protect the threat of spreading pests, contamination or diseases into the country of import or new place. The importer will require a phytosanitary certificate to import regulated material including commodities such as plants, bulks & tubers or seeds for propagation, fruits & vegetables, cut flowers & branches, grain and growing medium. They may also be required for certain plant products that have been processed and have a potential for introducing regulated pests (examples include cotton or wood). Sometimes they can also be required to cover contaminated materials such as empty shipping containers, vehicles or other organisms. If the transporter cannot provide a phytosanitary certificate to the importer, it's highly likely that the goods will not be cleared through customs which is not transported from the arrived region. The commodities may be seized and completely destroyed. Phytosanitary certificates can be in paper form or, where it is accepted by the NPPO of the importing country, in electronic or by fax.

Pest Risk Analysis (PRA) in Plant Quarantine

- Initiation
- Pest Risk Assessment
- Pest Risk management

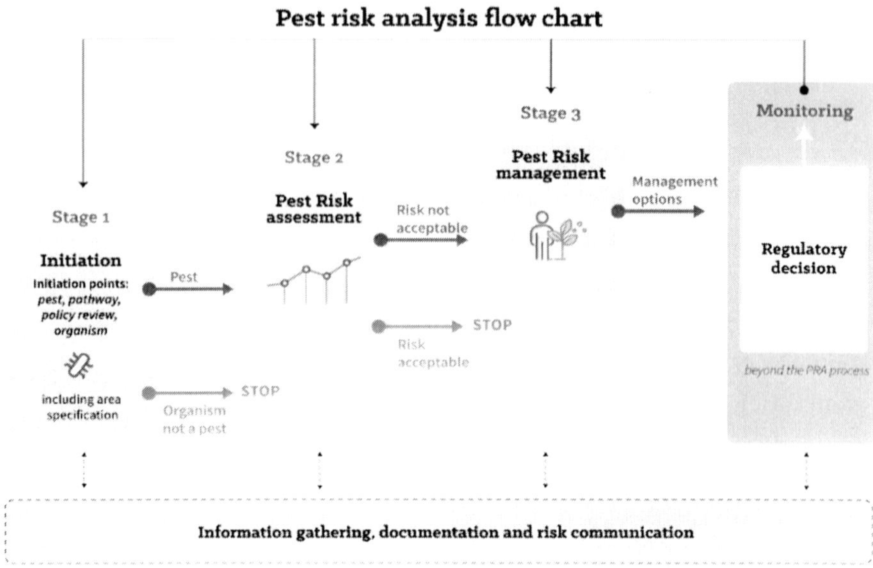

Presently, there are 78 Plant quarantine Stations (2023) at located in different location like Airports, Seaports and Land Frontiers.The regulations of plant quarantine station by Headquarters which located in Faridabad.

Regional Plant Quarantine Station

1. RPQS, Sewari (East), Mumbai
2. PQS, Air Cargo, Mumbai
3. PQS, JNPT, Nhava Sheva, Mumbai
4. Navkar Unit, Mumbai
5. PQS, Nashik
6. PQS, Nagpur
7. PQS, Goa
8. PQS, Indore
9. PQS, Raipur

Problems in Plant Quarantine

- Don't detect the all type of infection of insect pests and pathogens by conventional method.
- Lack the organized services, institute and other quarantine at state level.
- Lack of inter-state border PQ check posts and railway and road links which are transfera the insect and microorganism with the plant product.

- The method may not be sensitive enough and hidden infestation to reveals traces of infection. Which are not control the spread of insect pest and disease.
- The latent infectious may pass or not justified so undefected under post entry quarantine.
- Destruction of all infected or suspected material. Which are more chance to spread of insect pest and disease.
- Some exporter are apply the chemical for not germinate the microorganism so use sensitive method which are lack for testing insecticides treated seeds.

Agencies in involved in plant quarantine in India

Directorate of Plant Protection, Quarantine and Storage (DPPQS) Authority are employ the quarantine rules and regulation, recommendation, organization handles, mass import and export of seed and plan product for commercial purposes.

ICAR- National Bureau of Plant Genetic Resources (NBPGR) New Delhi National nodal agency for exchange of germplasm material of Agriculture crops for research purposes in the country. Quarantine responsibilities important to the germplasm. The Director of NBPGR has been empowered to issue "Permission" for imports of seeds, fruit and planting materials for Research purpose.

National Institute of Plant Health Management Premier Autonomous Institution under the Department of Agriculture and Cooperation, under Government of India with the mandate (Defenders) to promote; environmentally based Plant Health Management practices in diverse and changing agro-climatic conditions and environment, Chemically Management and Biosecurity & Incursion Management through capacity building programmes; and extend policy support to Central and State Governments.

Plant Biosecurity Division is involved in capacity building in various aspects of quarantine. Which are based on the biological and ecological parameter.

National Bureau of Agricultural Insect Resources (NBAIR): It maintains database with respect to invasive pests, nonnative pest and potential invasives. Engaged in Biosecurity, threat perception with action-plan for alien pest.

Role of NBPGR in Plant Quarantine: To supress the introduction of exotic pests or their new virulent races/strains into new areas. Several examples of dangerous pests introduced along with plants, plant products that have led to serious socio-economic consequences. Plant quarantine is a statutory requirement to regulate the entry of seed/planting material, plant products, living organisms and soil etc.to prevent highy movement pathogens entry of pests and pathogens. International exchange of plant genetic resources (PGR) is important to broaden the genetic base of crops in order to develop improved crop varieties. The Government of India has legislated the PQO 2003 to regulate and supress the import of plant material. Under this Order, ICAR-(ICAR-NBPGR) has been generate the powers to issue Import Permit and to carry out quarantine processing of imported PGR including transgenics and modified and for issue of Phytosanitary Certificate for PGR meant for export.

International Standards, Trade and Policy

- World Trade Organization (WTO)
- MRL's Established by CODEX
- Sanitary and Phytosanitary measures (SPS)
- Phytosanitary Certification (PC)
- Pest Risk Analysis (PRA)
- North American Plant Protection Organization (NAPPO)
- International Plant Protection Convention (IPPC)

National Standards

- Post Entry Quarantine (PEQ)

Conclusion

PQ is regulatory program of managing and suppress the insect pests, disease and weeds associated with movement of plant/plant product between two countries and also between state or with in country.

PEQ is accomplished through prohibition, interception, elimination and eradication of pests, diseases and weeds at post entry point. An effective implementation ofPEQ can management in India without risking environmental safety, economically, and ecological balance.Chemically based postharvest treatments are being replaced by biological and physical treatments and systems approaches. Drawing upon a wider knowledge base, by assessing risks in a more interdisciplinary fashion and including stakeholders from an inspection at early stage or before the regulations and actions should become more sustainable.Current implementation of quarantine is greatly emphasized for managing of pests, which in turn helps in conserving the productivity of crops and movement of pathogen from foreign countries.

References

Akhtar, J., Singh, B., Kumar, A.K.P., Maurya, A.K. and Dubey, S.C. (2017). Interception of pathogens during quarantine processing: an effort towards safe import of oilseed and vegetable Brassicas germplasm in India. Journal of Oilseed Brassica, 81(2): 120-130

Bhalla, S. and Gupta, K. (2016). Biosecuring India against exotic bruchids: two decades of quarantine monitoring of grain legumes germplasm. Indian Journal of Entomology, 78 (special): 79-90

Bhalla, S., Chalam, V.C., Singh, B., Gupta, K. and Dubey, S.C. (2018) Biosecuring plant genetic resources in India: role of plant quarantine. ICAR-National Bureau of Plant Genetic Resources, New Delhi, p 216

Chand, P., Singh, A., Vishwakarma, R. and Singh, C.K. (2017). Plant quarantine: an effective approach for prevention of alien pest and disease. Bull. Env. Pharmacol. Life Sci., 6(11): 07-12

Dubey, S.C. and Gupta. K. (2016). Plant quarantine system for PGR in India.Indian J. Plant Gene. Resour., 29: 410-413

Dubey, S.C., Gupta, K., Akhtar, J., Chalam, V.C., Singh, M.C., Khan, Z., Singh S.P., Kumar, P., Gawade, B.H., Kiran, R., Boopathi, T. and Kumari, P. (2021). Plant quarantine for biosecurity during transboundary movement of plant genetic resources. Indian Phytopathology, 74(2): 495-508

FAO Glossary of Phytosanitary Terms (1990). International Plant Protection Convention, 1992. FAO, Rome, FAO Plant Protection Bulletin 38(1): 5-23

Laxmi, V.R. Geetanjaly and Sharma, P. (2014). Plant quarantine: an effective strategy of pest management in India. Research Journal of Agriculture and Forestry Sciences, 2(1): 11-16

Luvisi, A., Nicolì, F., and De Bellis, L. (2017). Sustainable management of plant quarantine pests: The case of olive quick decline syndrome. Sustainability, 9(4): 65-69

Martin, R.R., Constable, F. and Tzanetakis, I.E. (2016). Quarantine regulations and the impact of modern detection methods. The Annual Review of Phytopathology, 4(54): 189-205

Singh, B., Akhtar, J., Kandan, A., Kumar, P., Chand, D., Maurya, A. K., Agrawal, P. C. and Dubey, S.C. (2018). Risk of pathogens associated with plant germplasm imported into India from various countries. Indian Phytopathology, 71(1): 91-102

16

Agrochemicals and Their Impact on Environment

Mukesh Kumar Mishra, Shailendra Kumar Mishra, Lakshmi Panday[3] and Rakesh Pandey

Department of Entomology Banda University of Agriculture & Technology, Banda, Uttar Pradesh, India

Introduction

An agrochemical is a chemical product used in industrial agriculture. In most of the cases, agrochemicals refer to pesticides (insecticides, herbicides, fungicides, algaecides, rodenticides, molluscicides and nematicides), fertilizers, soil conditioners, liming and acidifying agents and plant growth regulators that provide benefits and manage agricultural ecosystem. Agrochemicals were introduced to enhance crop yields and minimized the crop losses due to pests and categories as fertilizers and pesticides. In the past sixty years with the wake of the green revolution (1960s), use of nitrogen, phosphorus and potassium based inorganic fertilizers and pesticides has greatly increased food production by improving efficient and economical crop production in order to fulfil the food needs of the rapidly expanding world population (Pal *et al.*, 2006). The history of pesticide use can be divided into three periods of time. During the first period before the 1870s, pests were controlled by using various natural compounds. During the second period, between 1870 and 1945, people began to use inorganic synthetic materials. At the end of the 1800s, people in Sweden used copper and sulfur compounds against fungal attack in fruit and potatoes. Since then, people have been using many inorganic chemicals, including the Bordeaux mixture, based on copper sulfate and lime arsenic, as pesticides. The third period started after 1945, represented by the use of synthetic pesticides with the discovery of the effects of DDT, BHC, Aldrin, Dieldrin, Endrin, Chlordane, Parathion, Captan and 2,4-D. After the entry of pesticides in agricultural ecosystem it become an integral part of agriculture and the research has continued into producing more selective pesticides. The pest, disease and weeds can loss more than half of our crops without pesticides, thus the pesticides enable farmers to produce safe, quantity & quality foods at affordable prices. According to Food and Agriculture Organization, the world highest pesticide consumption (13.35 kg active ingredient ha^{-1}) was recorded in Taiwan followed by Japan (11.89 kg a.i.). In India, the present pesticide consumption is around 337g ha^{-1} (technical grade) which is lower than the world average of 645 g ha^{-1} and other developing countries. During 2021-22, the total cultivated area of India was about 173.99 mha and only 20.96% of that was devoid of any pesticide application. As per the record of Central Insecticide Board and Registration Committee (CIB&RC) updated in October 2022, a total of 318 pesticides have been registered in India for use in agriculture against various pests and diseases. These agrochemicals and pesticides are expensive and polluting (Bahadur *et al.*, 2015). As per current research, the global agrochemicals and pesticides market was estimated at 278428.78 USD million as of 2023 and it's anticipated to reach USD 320916.55 million by 2028, with a CAGR of 2.4% during the forecast years (Anonymous, 2023).

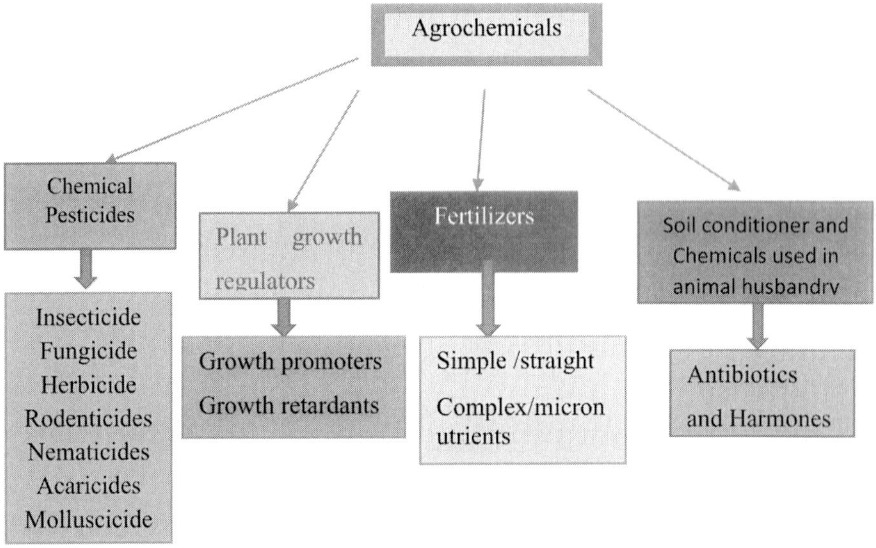

Fig. 1. Broad classification of agrochemicals

The over and misuse of agrochemical and pesticides has precedence to immense health problem, economic loss and various environmental problem and had an impact on soil microorganisms as well as the entire ecosystem (Onder *et al.*, 2011). The overuse of agrochemicals also leads to environmental pollution such as air, water, soil pollution, etc. and cause imbalance in the ecosystem and it pollute water resources by getting into bodies of water (Aktar *et al.*, 2009). The primary decreasing biotic factors in agriculture are weeds and insects, which reduce crop production, productivity, and resource use effectiveness (Oliveira *et al.*, 2014). Besides, synthetic fertilizers and pesticides helped by increasing agricultural yield and shield crops from disease pathogens and phytophagous pests, respectively. In general, the dosage of the chemical, the characteristics of the soil, and many environmental conditions will affect the effects of pesticides on microorganisms differently.

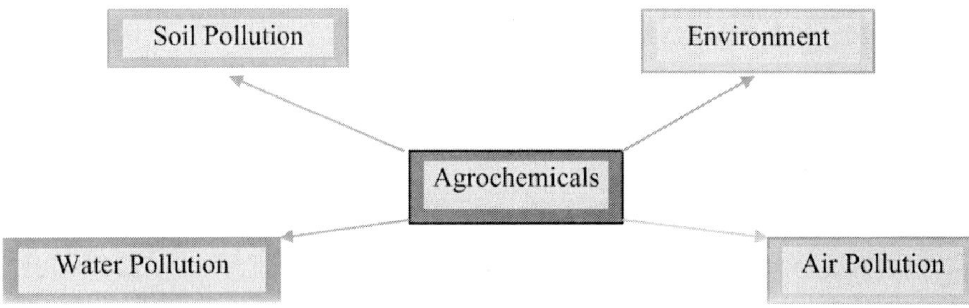

Fig. 2. Impact of agrochemicals on environment

Agrochemicals and their effects

The persistent and careless use of agrochemicals has made various issues *viz.*, deterioration of soil sterility, pest resurgence, destruction of pollinators and natural enemies, pesticide resistance, drug-resistant phytopathogens, human and animal health hazard, diseases and disorders of human limbs, destruction of native flora and fauna, issues related to bioaccumulation and bio magnification, climate change, environmental pollution etc. worse, posing an alarmingly serious threat to life and health. Unbalanced nutrient levels and decreased crop yielding capability are results of overusing these compounds, which also produce residues. According to Han and Farooq (2000), a severe problem arises when bacteria develop a resistance to pesticides and fish in various bodies of water have died due to pesticide usage.

1. Effect on environments

A healthy environment is essential for optimum human health and contamination of soil and water through human and industrial waste and agrochemicals is a universal problem and fact in developing countries. The harmful effects of agrochemicals on the environment are brought on by its chemical resources, which also appear to have a detrimental influence on people, animal and whole nature. Unrestrained domestic and industrial waste in urban and rural situations and pollution of reservoirs caused by agrochemical and pesticides runoff are an increasing concern. Deforestation and consequent soil erosion further aggravate these issues. Misuse of agrochemicals is a key cause of pollution of potable water, reservoirs, and watersheds and it is mostly attributable to irresponsible behaviour of farmers and agrochemical sellers and large government fertiliser subsidies that incentivise fertilizer overuse are fully preventable (Wimalawansa and Wimalawansa, 2014). Ultimately those are persisting for long time in the environment and causes environmental pollution.

2. Effect of agrochemicals on soil

The overuse of synthetic fertilizers and pesticides impacted as decline in organic matter, the soil's micro biota and microbial enzymes (Kalia and Gosal, 2011).However, the usage of herbicides impacts on the rhizobium-legume symbiosis (Drew *et al.*, 2007) and a decrease in the population of soil fungus has caused due to application of fungicide (Kalia and Gosal, 2011). Besides, the agrochemicals may also kill beneficial soil micro-organisms, increase nitrate level in soil, alter the soil pH, reduced soil quality by increasing residual effect and toxicity.

a) Impact of agrochemicals on soil microorganism

The soil micro flora plays a crucial role in the process of carbamate organic matter decomposition. The impact of pesticides on soil micro flora growth, activity, and enzyme levels may lead to negative consequences for soil fertility and overall soil health. The study conducted by Ghoshal and Hati (2019) studied that the utilization of Rynaxypyr, cartap hydrochloride, fipronil, and chlorpyrifos did not have any discernible adverse impacts on the population of collembolan. However, the application of carbofuran and phorate resulted in a reduction of 27.65% and 13.47%, respectively, in the rice-maize cropping system. Soil contamination or soil pollution is caused by the presence of xenobiotic (human-made) chemicals or other alteration in the natural soil environment. Industrial activity, agricultural chemicals or improper disposal of waste are on the hit-chart responsible for land degradation (Pangos *et al.*, 2013).

b) Impact of agrochemicals on soil microbial metabolism

Agrochemicals also interfere with microbial metabolism and change soil enzyme activity. Agrochemicals in soil are biodegraded in part by soil enzymes. Soil enzymatic activity is a significant bio indicator of soil health and fertility due to its function in maintaining soil equilibrium (Mandal *et al.*, 2020 and Virág D *et al.*, 2007). Herbicides, atrazine, and metolachlor, in particular, have a negative impact on the invertase and dehydrogenase enzymes' abilities to function. Fungicides including benomyl, mancozeb, thiram, and tridemorph block the action of phosphatase, urease, and dehydrogenase. Agrochemicals, particularly pesticides meant to kill certain target soil microorganisms, also cause nontarget microorganisms to multiply quickly and thrive.

3. Effects of agrochemicals on water

Agrochemicals pass into the water bodies by different processes such as runoff, drift and leaching via the soil or they may be applied directly into surface water to control some aquatic pests or mosquitoes. Insecticides and herbicides can affect fish and invertebrates and alter the species composition of aquatic commodities by mixing in surface water through spray-drift or run-off. About 41% of all amphibian species are threatened (IUCN 2020). According to IUCN, 2009, one-third of 6000 amphibian species worldwide are threatened. Besides habitat loss, overexploitation, or introduced species, amphibians are affected by the pollution of surface waters with fertilizers and pesticides (Praween and Jan, 2020). Due to impact of agrochemical water becomes unfit for consumption, it promotes the growth of algae and it leads to eutrophication due to excess use of chemicals fertilizers. Aquatic animals are exposed to agrochemical or pesticides in three ways namely, breathing (uptake via gills during breathing), dermally (direct absorption via skin), and orally (entry via drinking contaminated water). It can affect aquatic forms by decreasing dissolved oxygen in the water and can trigger physiological and behavioural changes in fish populations. A study in India revealed the presence of Organochlorine pesticides in 58% of drinking water samples which were taken from various hand pumps and wells around Bhopal.

4. Effects of agrochemicals on Air

The quality measurement of air is based on the amount of pollutants in our atmosphere, including indoor and outdoor air. Agrochemicals particularly, pesticides in agriculture and urban settings have the potential to contaminate our air, affecting human, animal and plant health. Some pesticide ingredients stay in the atmosphere for short time period, while others can last longer. Agrochemical released into the air can settle to the ground, be broken down by sun light and water in the atmosphere, diffuse with air and altering their composition. Residues and particles of agrochemicals can lead to air pollution. Pesticides drift may be happens when air carries the agrochemicals particle from one place to another place and more amount of spray evaporation may be happen due to low relative humidity and higher temperatures. It can cause respiratory illness on surrounding organisms' health due to inhalation of polluted air by the use of agrochemicals.

5. Impact of agrochemicals on non-target flora

Herbicides have an impact on the variety of flora (Gove *et al.,* 2007). The use of herbicides has a greater impact on young plants. When plants are exposed to the highest concentrations of herbicides, their reproductive systems are impacted (Boutin *et al.,* 2014). Beginning with the second year of exposure, the spray of agrochemicals drift impacts the balance of species. Additionally, this disrupts the diversity of plant species. Fungicides and insecticides have a negative impact on non-target flora. Plant photosynthesis is impacted by the application of fungicides (Petit *et al.,* 2008). The metabolism of both carbon and nitrogen is also impacted (Saladin *et al.,* 2003). A decrease in floral visits and a change in the community of floral visitors both have an impact on pollination effectiveness (Cahill *et al.,* 2008). According to Jennings *et al.* (2012), pesticides also have an impact on plant development and germination.

6. Impact of agrochemicals on vertebrates

Organophosphates and carbamates impair vertebrates' ability to regulate body temperature, consume nutrients, and forage for food, reproduce, and other metabolic processes (Story and Cox, 2001). In addition to affecting the thyroid function in rodents, birds, amphibians, and fish (Brucker-Davis, 1998), insecticides, herbicides, and fungicides also have an impact on the immune system (Galloway and Handy, 2003). According to Sanchez-Bayo (2012), terrestrial vertebrates are less vulnerable to pesticides made from plant or fungal toxins than are birds and mammals.

7. Effect of agrochemicals on snails and slugs

Snails and slugs exhibit the capability to accumulate insecticides, specifically organophosphates and carbamates, within their anatomical structures. The present study revealed that the organisms under examination contained elevated levels of diazinon, phorate, and carbofuran, which can be attributed to their water solubility. The usage of insecticides might not pose any detrimental effects on the intended target, but it is deemed harmful towards the predatory avian species that subsist on the snails and slugs. These avian creatures are at severe risk of perishing as a result of the utilization of insecticides. Agrochemicals Chemical pesticides Plant growth regulators Fertilizers Soil conditioner and Chemicals used in animal husbandry.

8. Association between agrochemicals and climate change

A consequence of climate change is that agrochemical usage has grown as a result of an increase in pest and disease outbreaks brought on by warmer temperatures. Climate issues are also brought on by the production and application of agrochemicals to combat pests and illnesses. Fuel from fossil fuels must be burned in order to produce nitrogen-based fertilisers, which is an energy-intensive process. Chains of chemical reactions produce nitrous oxide, or N_2O, a greenhouse gas, after farmers apply these synthetic fertilisers to crops. Compared to carbon dioxide and methane, N_2O has a 265-times-greater potential to cause global warming. (https://www.bloombergquint.com/onweb/synthetic-fertilizer-ammonium-nitrate-makes-climate-change-worse).

Management strategies of agrochemicals to minimise their effect

It's crucial to choose safe technologies for the use of agrochemical. The danger effect of agrochemicals can be minimized by through the use of safe working systems. The management strategies includes:

- Users of agrochemicals should keep in mind to ask for advice when necessary.
- Selection of pesticides based on its lowest toxicity to humans, mammals, fish, birds and other invertebrates.
- Use less toxic agrochemical with higher soil absorption, lower water solubility, shorter half-life, most persistent and have greater chance of off-site movement via runoff.
- There may be several varieties of spray equipment on the market, but the greatest ones are frequently those that are ideal for safety.
- Agrochemical should have, low runoff potential e.g. liquid applications rather than granular formulations of pesticides.
- It should science based holistic environmental management approaches aimed at reducing the possibility of pesticides and other chemicals and sediments of moving off treated areas into receiving water bodies.
- Use of recommended doses of agrochemical Control the rate, method timing and type of chemicals being applied.

Conclusion

Agrochemicals are regarded as a potent tool or "magic bullet" in underdeveloped nations for increasing agricultural output and significantly raising key public health indicators. However, it has been shown that agrochemicals pose significant risks. The ecology is negatively impacted by pesticide use. The employment of biological techniques must be an option to using agrochemicals. The growth of the plants is impacted by agrochemicals. The use of agrochemicals results in the death of insects, pests, nematodes, weeds, etc., but they are also environmentally unfriendly and polluting. It is necessary to thoroughly investigate their molecular level mechanism.

References

Aktar W., Sengupta D., Chowdhury A. (2009). Impact of pesticides use in agriculture: Their benefits and hazards. Interdisci. Toxicol. 2: 1-12

Anonymous (2023). https://www.marketwatch.com/press-release/latest-2023-agrochemicals-and-pesticides-market-overview-forecast-by-2030-with-104-pages-2023-05-22

Bahadur S., Verma S.K., Prasad S.K., Madane A.J., Maurya S.P., Verma V.K., Sihag S.K. (2015). Eco-friendly weed management for sustainable crop production - A review. J. Crop Weed, 11: 181-189

Biswas S.K., Rahman S., Kobir S.A., Ferdous T., Banu N.A. (2014). A review on impact of agrochemicals on human health and environment: Bangladesh perspective. Plant Environ. Develop. 3: 31-35

Boutin C., Strandberg B., Carpenter D., Mathiassen S.K., Thomas P.J. (2014). Herbicide impact on non-target plant reproduction: What are the toxicological and ecological implications? Environ. Pollut., 185: 295-306

Brucker-Davis, F. (1998). Effects of environmental synthetic chemicals on thyroid function. Thyroid, 8: 827-856

Cahill J.J., Elle E., Smith G.R., Shore B.H. (2008). Disruption of a belowground mutualism alters interactions between plants and their floral visitors. Ecology, 89: 1791-1801

Donaldson D., Kiely T., Grube A. Pesticide's industry sales and usage 1998-1999 market estimates. US Environmental Protection Agency; Washington (DC): Report No. EPA-733-R-02-OOI. Available from: http: //www.epa.gov/oppbead/ pesticides/99 pestsales/market-estimates.pdf. [Google Scholar]

Drew E.A., Gupta V.R., Roget D.K. (2007). Herbicide use, productivity, and nitrogen fixation in field pea (*Pisum sativum*). Aus J Agr Res, 58: 1024-1214

Galloway T, Handy R. 2003. Immunotoxicity of organophosphorous pesticides. Ecotoxicol, 12: 345-363

Ghosal A., Hati A. (2019). Impact of some new generation insecticides on soil arthropods in rice maize cropping system. The Journal of Basic and Applied Zoology, 80(6): 1-8

Gove B., Power S., Buckley G., Ghazoul J. (2007). Effects of herbicide spray drift and fertilizer overspread on selected species of woodland ground flora: comparison between short-term and long-term impact assessments and field surveys. J. Appl. Ecol., 44: 374-384

IUCN (2020). The IUCN red list of threatened species. Version 2020-1.https://www.iucnredlist.org. Accessed 2 June 2020

Jennings D.E., Congelosi A.M. and Rohr J.R. (2012). Insecticides reduce survival and the expression of traits associated with carnivory of carnivorous plants. Ecotoxicol, 21: 569-575

Kalia A., Gosal S.K. (2011). Effect of pesticide application on soil microorganisms. Arch Agron Soil Sci, 57: 569-596

Mandal A., Sarkar B., Mandal S., Vithanage M., Patra A.K., Manna M.C. (2020). Impact of agrochemicals on soil health. In Agrochemicals Detection, Treatment and Remediation, ed. M Narasimha, V Prasad, pp. 161–87. Oxford, UK: Butterworth-Heinemann

Oliveira C.M., Auad A.M., Mendes S.M., Frizzas M.R. (2014). Crop losses and the economic impact of insect pests on Brazilian agriculture. Crop Prot, 56: 50-54

Onder M., Ceyhan E., Kahraman A. (2011). Effects of agricultural practices on environment. Biol Environ Chem, 24: 28-32

Pal, R., Chakrabarti, K., Chakraborty, A., Chowdhury, A. (2006). Degradation and effects of pesticides on soil microbiological parameters—a review. Int. J. Agric. Res. 1(33): 240-258

Panagos P. Van Liedekerke M., Yigini Y, Montanarella L. (2013). Contaminated Sites in Europe: Review of the current situation based on data collected through a European Network. Journal of Environmental and Public Health. 10: 1155/2013/158764

Parween, T. and Jan, S. (2020). Pesticide consumption and threats to biodiversity in Ecophysiology of Pesticides:Interface between Pesticide Chemistry and Plant Physiology Pages 39-73

Petit A.N., Fontaine F., Clément C., Vaillant-Gaveau N. (2008). Photosynthesis limitations of grapevine after treatment with the fungicide fludioxonil. J Agri Food Chemi, 56: 6761-6767

Saladin G., Magne C., Clement C. (2003). Effects of fludioxonil and pyrimethanil, two fungicides used against Botrytis cinerea, on carbohydrate physiology in Vitis vinifera L. Pest Manage Sci, 59: 1083-1092

Sanchez-Bayo F. (2012). Insecticides mode of action in relation to their toxicity to non-target organisms. J Environ Anal Toxicol, S: 4: 2

Story P., Cox M. (2001). Review of the effects of organophosphorus and carbamate insecticide vertebrates. Are there implications for locust management in Australia? Wildlife Res, 28: 179-193

Virág D., Naár Z., Kiss A. (2007). Microbial toxicity of pesticide derivatives produced with UV photodegradation. Bull. Environ. Contam. Toxicol. 79(3): 356–59

Wimalawansa, Shehani A. and Wimalawansa, Sunil J. (2014). Agrochemical-Related Environmental Pollution: Effects on Human Health. Global Journal of Biology, Agriculture and Health Sciences, 3(3): 72-83

17

Pesticides and Their Formulations

***Mukesh Kumar Mishra*[1], *Shailendra Kumar Mishra*[2]**
***Lakshmi Pandey*[3] *and Arun Kumar*[4]**

[1,2&3]*Department of Entomology, Banda University of Agriculture and Technology Banda-210 001, Uttar Pradesh, India*

[4]*Department of Entomology, Chandra Shekhar Azad University of Agriculture and Technology, Uttar Pradesh, India*

Introduction

The toxicant must be able to be applied effectively in order to come into touch with the pest or leaf and to leave a uniform and permanent deposit on the plant surface. Pesticide in chemically pure form is known as technical grade pesticides. These cannot be applied as such due to high toxicity and complex physical properties. They are converted into a ready to-use form known as formulation. Formulation converts the technical grade pesticide into a product which can be suitably diluted & applied by practical methods. Which have homogeneous distribution over large area, convenient to handle and more attractive to the pest. Formulation can be defined as "a physical mixture of pesticide with inert ingredients which can be suitably stored, transported and applied by practical means to achieve effective, safe and economic control of pests" or it is a substance which processed through formulation in order to increase its storage, handling, application, efficacy, safety for the user and the environment, and financial success. Grewal (2005) defined as "formulation is the process of creating a product from an active component by adding various active (functional) and non-active (inert) ingredients". or the formulants are defined as inert diluents, they are usually not required to be tested for chronic toxicity (R. Mesnage *et al.*, 2013 & 2019). It is the last physical state in which pesticide is offered for sale. The mode of application is generally constrained by the formulation, which describes what and how other ingredients are combined with the active component.

There are several "formulas" for pesticides and majority of technical pesticides are created by combining one or more active ingredients (a.i.), which kill pests with inert substances, diluents, preservatives, adjuvants, etc. prior to use in order to create a product that is efficient, simple to handle and apply, has a good shelf life, and is free of unfavourable side effects. The qualities of the inert component, coupled with the properties of the active ingredient, define the signal word that is shown on the product label. Some inert substances offer health concerns to pesticide handlers or applicators. Due to the inert ingredients present in the formulation, pesticide products with the same active ingredient are marketed in the form of granules, dusts, solutions, wettable powder, flowable suspension concentrate, emulsifiable concentrate, spray able oils, aerosols etc. (Muntz *et al.*, 2016).

A formulation is categorised as based on their physical condition in the container at the time, solids or liquids buy time. More than one active component may be included in a formulation, and many of them must be further diluted with a suitable carrier (such as water) before usage. Several alternative formulations of the same pesticide are frequently offered for sale and it is grouped in two major categories:

- **Conventional Formulations**: Formulations which are being used in agriculture and public health sector for long time viz., Dustable Powder (DP), Wettable Powder (WP), Emulsifiable Concentrate (EC), Soluble Concentrate (SL), Granules (GR): Coated & Impregnated and Baits formulation.
- **Safer Formulation or New Generation Formulations:** These are safe, user & environment friendly effective products viz., Suspension Concentrate (SC), Water Dispersible Granules (WG), Concentrated Emulsion (EW), Microemulsion (ME), Controlled Release Formulation (CR), Suspo-emulsion Formulation (SE), Tablet Formulation (WT) and Gel Bait.

The pesticide formulations are also grouped on the basis of their physical appearance in the market as solid, liquid and other formulations:

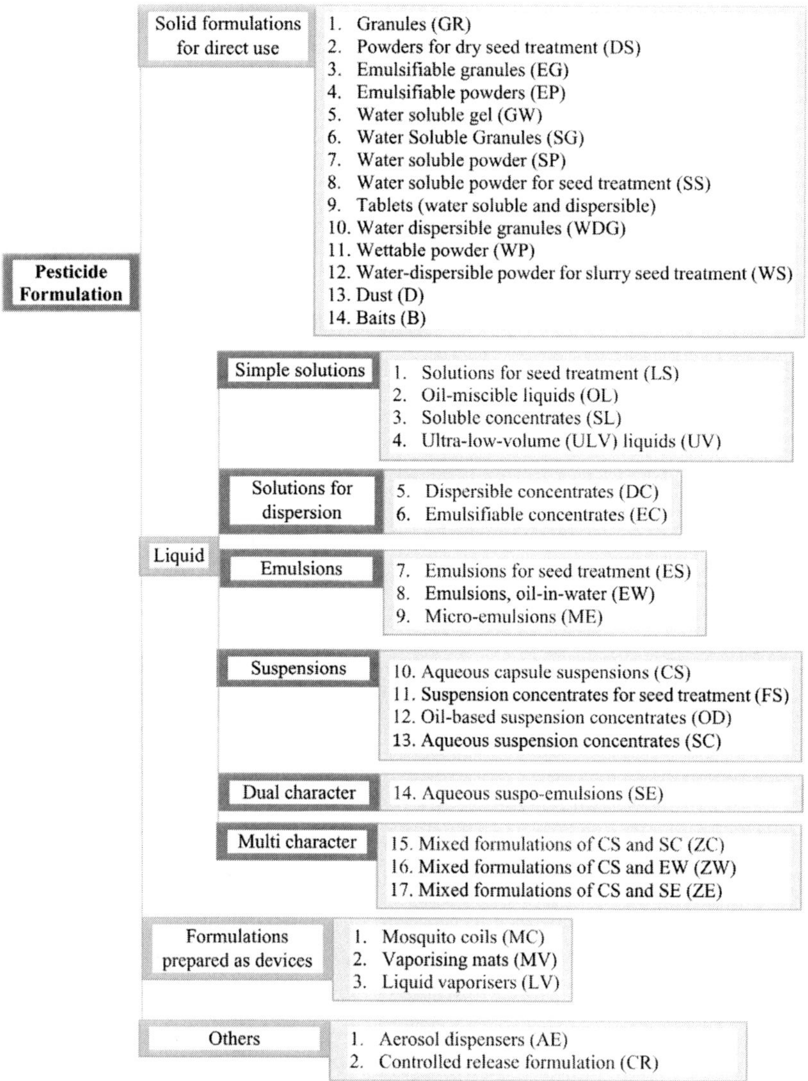

Fig. 1. Classification of pesticidal formulations (Source: http://apvma.gov.au/node/10901)

Solid formulation

1. **Granules (G): (Ingredient**: Pesticide, carriers, (river sand/marble hips), Coating agent) Formulations that are ready to use are granules. Either the granules' outside is coated with the active component, or it is absorbed by tiny clay, talc, or other carrier particles. The amount of
2. **Powders for dry seed treatment (DS):** This type of formulation look like a powder for application in the dry state, directly to the seed.

3. **Emulsifiable granules (EG):** It is a granular formulation, which may contain water-insoluble formulants, to be applied as an oil-in-water emulsion of the active constituent(s) after disintegration in water.
4. **Emulsifiable powders (EP):** It is a powder formulation, which may contain water-insoluble formulants, to be applied as an oil-in-water emulsion of the active constituent(s) after dispersion in water.
5. **Water soluble gel (GW):** It is a gelatinized formulation to be applied as an aqueous solution.
6. **Water Soluble Granules (SG):** It is a granule formulation to be applied as a true solution of the active ingredient after dissolution in water, but which may contain insoluble inert ingredients.
7. **Water soluble powder (SP):** A powder preparation to be applied as a true solution of the active substance after solution in water but which may contain insoluble. These mixtures resemble wettable granules in appearance. But when combined with water, soluble powders easily dissolve and create a real solution. There is no need for extra agitation after a thorough mixing. Soluble powders include an active component in concentrations ranging from 15% to 95%.
8. **Water soluble powder for seed treatment (SS):** It is a powder formulation to be dissolved in water before application to the seed.
9. **Tablet Formulation** (**Main components**: effervescent tablet includes active ingredients, effervescent disintegrants, thinners, adhesives, surfactants, lubricants, glidants and others.

   It is characterized by low or no water solubility, a complementary delivery system containing an organic acid, an inorganic base, a dispersant, a disintegrant, and a wetting agent an contain about 20% to 75% of a pesticide.

   a) **Water soluble tablets (ST):** It is a tablet formulation to be used individually, to form a solution of the active ingredient after disintegration in water. The formulation may contain water-insoluble formulants.

   b) **Water-dispersible tablets (WT):** It is a tablet formulation to be used individually, to form a dispersion of the active constituent after disintegration in water.
10. **Water dispersible granules (WDG):** These are granule-sized, dust-free particles made from compressed WP formulations. The granules disintegrate into a fine powder when they are stirred in water. These products contain wetting agent and dispersing agent similar to those

used in wettable powders, but the dispersing agent is usually at higher concentration. Water dispersible granules are usually more expensive than older types of formulations (dusts, wettable powders) but their safety and greater convenience regarding application make them still desirable for many users (Knowles, 2008).

11. **Wettable powder (WP):** (Ingredient: pesticide, diluents/ carrier, wetting and dispersing agents) These are the pesticide formulation consisting of the active ingredient in a finely ground state combined with wetting agents and sometimes bulking agents (www.crodacropcare.com). Wettable powder are free-flowing powders that contain wetting and dispersion agents, carriers, and a majority of technical material between 25 and 75 percent. When dissolved in water, WPs create suspension.
12. **Water-dispersible powder for slurry seed treatment (WS):** It is a powder formulation to be dispersed at high concentration in water before application as slurry to the seed.
13. **Dust (D): (Ingredients:** Pesticide, Carrier/Diluents): It is a pesticide formulation in which, the toxicant is diluted either by mixing with or by impregnation on a suitable finely divided carrier. The finished product may contain 0.1-25 per cent of the toxicant. The particle size of dusts ranges between 1 and 40. In general, the toxicity of an insecticide increases as the particle size decreases.
14. **Baits (B): (Composition**: Pesticide, Diluents, food attractant): These are ready-to-use formulations (RTUs) that combine an active component with food or another alluring material. Pests are either drawn to the bait or it is placed in an area where they will locate it.

Liquid formulation

A. Simple solution

1. Solutions for seed treatment (LS): It is a clear to opalescent liquid to be applied to the seed either directly or as a solution of the active constituent after dilution in water. The liquid may contain water-insoluble formulants.
2. Oil-miscible liquids (OL): It is a homogeneous liquid formulation to be applied as a homogeneous liquid after dilution in an organic liquid.
3. Soluble concentrates (SL): It a clear to opalescent liquid to be applied as a solution of the active constituent after dilution in water. The liquid may contain water-insoluble formulants.
4. Ultra-low-volume (ULV) liquids (UV): It is a homogeneous liquid ready for use through ultra-low-volume (ULV) sprayer. These products are

designed to be sprayed without dilution, and thus must be applied very carefully with specialized equipment. In addition to the active ingredient and a solvent system in which the active is very soluble, UL products often contain surface-active agents and drift control substances.

B. Solutions for dispersion

1. **Dispersible concentrates (DC):** It is a liquid homogeneous formulation to be applied as a solid dispersion after dilution in water.

2. **Emulsifiable concentrate (EC):** (Ingredient: Pesticide, Solvents, Emulsifiers, Stabilizers)The formulation consists of the toxin, the toxin's solvent, and an emulsifier. It is a transparent solution that, when diluted with water to spray strength, forms an emulsion of the oil-in-water kind. Spraying a solvent causes it to swiftly evaporate, leaving a toxicant deposit where water also evaporates.

C. Emulsions

1. **Emulsions for seed treatment (ES):** This formulation is a stable emulsion for application to the seed, either directly or after dilution.

2. **Emulsions, oil-in-water (EW):** (Composition: Active ingredient, emulsifier, antifreezing agent anti- foaming agent and preservative and water); This formulation is physically stabilised by specifically identified polymeric surfactants incorporated at an appropriate level. The emulsion has already been established in the formulation and is only diluted further in the spray mixture. They can be considered as a safer and more environmentally friendly alternative to emulsifiable concentrates. In an EW the continuous phase is water (as opposed to an organic solvent for ECs) which offers the benefit of lower phytotoxicity, no flashpoint concern, ease of handling, and a lower environmental impact.

3. **Microemulsion (ME)** (Composition: Active ingredient, Min. solvent Emulsifier, Co-surfactant water); Microemulsion is water-based formulations with a very small emulsified droplet size; this makes the formulation transparent. It is thermodynamically stable over a wide temperature range due to this very fine droplet size, usually between 0.01 and 0.05 μm.

D. Suspensions

1. **Aqueous capsule suspensions (CS):** This formulation is designation for a stable suspension of micro-encapsulated active ingredient in an aqueous continuous phase, which is normally intended for dilution with water before use.

2. **Flowable concentrates for seed treatment (FS):** It is a modification of suspension concentrates with supplemental additives for adhesion to the seed surface and colourants as safety markers to indicate that a seed has been treated with a product. It can be to the seed, either directly or after dilution and now the most popular type of seed treatment because they are concentrated formulations and are safer to apply as they are water based.
3. **Oil-based suspension concentrates (OD):** It is stable suspension of active constituent(s) in water-immiscible fluid, which may contain other, dissolved active constituent(s), intended for dilution with water before use. It is containing high melting a.i. with low solvent solubility and dispersed in oil as fine particles.
4. **Aqueous suspension concentrates (SC):** (Composition: Active ingredient, surfactants, thickener, antifreezing agent, antifoaming); The active component is included in this liquid formulation as a solvent that is compatible with water. During spraying, the solvent is combined with water and dissolves, leaving the active component intact. A surfactant is used to provide wetting power. They are created using a wet grinding technique, and their range of particle sizes is 1 to 10 m. Small particles are kept from re-aggregating during the grinding process by inert substances that are adsorbed onto particle surfaces. Due to the active ingredient's increased access to plant tissues due to the smaller particle size, these substances frequently display improved bio efficacy when used. They have several benefits since they are water-based, including ease of pouring and measuring, operator and environmental safety, and affordability. Consequently, they are growing in popularity as a formulation type (Woods, 2003; Knowles, 2005).

E. Dual character liquid formulation

1. **Aqueous suspo-emulsions (SE):** (Composition: Insoluble solid (a.i.) as dispersed particle, insoluble liquid (a.i.), as oil droplets, surfactants, water as bulk phase); It is a fluid, heterogeneous formulation consisting of a stable dispersion of active constituent(s) in the form of solid particles and fine globules in a continuous water phase. The formulation is intended for dilution into water prior to spray application.

F. Multi character liquid formulation

1. **Mixed formulations of CS and SC (ZC):** It is a mixed formulation of capsule suspensions and suspension concentrates. This formulation consists a stable aqueous suspension of microcapsules of one technical material combined with a fine particle of second technical material in an aqueous phase together with suitable formulants. In microcapsules, the

active ingredient is inside discrete, inert, polymeric microcapsules to provide slow release second active ingredient is free to provide immediate effect. This formulation is intended for dilution into water prior to spray application and it is easy to handle and measure, dust free, non-flammable and offer good miscibility with water. It is working as broad spectrum pesticide formulation.

2. **Mixed formulations of CS and EW (ZW):** It is a mixed formulation of capsule suspensions and emulsions, oil-in-water. The formulation consists of an emulsion of fine droplets of technical material in the form of the combination of a suspension of microcapsules in an aqueous phase together with suitable formulants. In the case of microcapsules, the active ingredient is present inside discrete, inert, polymeric microcapsules. After gentle agitation the material must appear homogeneous and be suitable for dilution in water worked and provide a broad-spectrum pest control.
3. **Mixed formulations of CS and SE (ZE):** It is a mixed formulation of capsule suspensions and suspo-emulsions. The formulation consists of an emulsion of fine droplets of technical material and a suspension of fine particles of technical material in an aqueous phase, together with suitable formulants. After gentle agitation the material must appear homogeneous and be suitable for dilution in water.

G. Formulations prepared as devices

1. **Mosquito coils (MC):** It a coil that burns without producing a flame and releases the active constituent into the local atmosphere as a vapour.
2. **Vaporising mats (MV):** It is a mat, which made from pulp, or other suitable inert materials, and impregnated with an active constituent. The mat is intended for use in a heating unit designed to produce slow volatilisation of the active constituent.
3. **Liquid vaporisers (LV):** It is liquid formulation in a cartridge, designed to fit a suitable heating unit, from which the formulation passes up a heated wick and evaporates into the local atmosphere.

H. Other formulations

1. **Aerosol dispensers (AE):** It consist of a liquid formulation in a pressurised, non-refillable aerosol dispenser, containing propellant(s), synergist(s) and other formulants, as required, intended for release of the active ingredient into the air in the form of an aerosol.
2. **Controlled release formulation (CR)** (Composition: Active ingredient Polymer, Surfactant) Controlled-release formulations are absorbed more slowly than conventional tablets and produce more stable serum levels

during the day and at night, even when given twice daily as compared with conventional tablets given three times daily. Peak serum concentrations are usually attained within 4–8 h with plain tablets.

3. **Neem based water dispersible granules:** It is a neem-based insecticide formulation with 0.1-3 mm diameter particles which are free flowing and dust free with a homogenous dispersion in water.
4. **Nanoemulsions of botanicals:** Many botanical based nanoemulsions *viz*., castor oil nanoemulsion, cashew nut shell liquid, neem nanoemulsion + biodiesel waste as co-solvent and neem oil nanoemulsion + botanical synergist are also available.

Nanopesticides: These are nanostructures with two to three dimensions between 1 to 200 nm, used to carry agrochemical ingredients (AcI) and synthesized by using carbon nanotubes and graphene-based nanoparticles. In formulation development and drug delivery, the term 'nanoparticles' refers to a variety of particles used to deliver drug products or proteins to their target site. These include liposomes, solid lipid nanoparticles (SLNs) micelles and polymeric nanoparticles. it increases the solubility of water-insoluble pesticides, attacks only on target pests, and protects the premature degradation of pesticide. The common nanocarrier types include polymers (b1–b4) such as chitosan, cellulose and polyethylene existing in the forms of nanocapsules (b1), nanospheres (b2), nano(hydro)gels (b3) and nanomicelles (b4), clay NMs (for example, silica, montmorillonite and kaolinite; b5), nanocomposites (b6), carbon nanotubes (CNTs; b7) etc. (Rajna *et al.,* 2019).

Purposes of pesticidal formulation

Formulations improves the properties of a chemical for handling, storage, application and may substantially influence effectiveness and safety (Knowles, 1998).

The following purposes of pesticidal formulations are mentioned below:

a) To make improve handling and application of pesticide easy

b) To maximize biological efficacy

c) To improve defects in pesticides

d) To make pesticides safer for workers and users

e) To reduce harmful effects and impact on non-target organism and the environments

f) To give various order to pesticides in order to broaden their spectrum of activity

Conclusion

Pesticide development, usage, disposal, and storage continue to be issues. Despite the fact that contemporary agrochemical businesses are dedicated to creating secure formulations to prevent persistent and bio accumulative. Given the inherent properties of chemicals, it is difficult to anticipate and prevent all potential detrimental effects of pesticides on human health and entire ecosystems. Even contemporary pesticides with purportedly short half-lives can linger in soil for up to many years in some circumstances, posing a risk to soil invertebrates. Alternative approaches (such as the creation of bio pesticides) might reduce harmful qualities and hasten the emergence of resistance in the quest of safety. To decrease the amount of dangerous ingredients in pesticide products and innovative technology are actively being developed. For instance, Nano carriers lessen pesticide losses and delay release, due to improved affinity to the target pest species, and increase the precision of delivery of active chemicals. Government regulations are a crucial part in ensuring the safety of agricultural goods for consumers. The testing of formulants' safety for non-target species based on thorough and trustworthy data on their effects, the necessary examination of the long-term impacts of adjuvants, and improving the predictive capacity of formulants.

References

Anonymous (2023). Formulation types. Australian pesticides and veterinary medicine authority. http://www.crodacropcare.com

Grewal, P.S. (2005). Formulations and Quality control of entomopathogenic nematodes. Department of entomology, Ohio-state-university Wooster

H. Muntz, R. Miller, D. Alston. (2016). Understanding Pesticide Risks: Toxicity and Formulation, (accessed March 24, 2021), http://npic.orst.edu/factshee ts/formulations.pdf

Knowles, A. (2008). Recent developments of safer formulations of agrochemicals. Environmentalist, 28(1): 35-44

Knowles, A. (2005). New developments in crop protection product formulation. Agrow Reports UK: T and F Informa UK Ltd., pp.153-156

Knowles, D.A. (1998). Chemistry and technology of agricultural formulations. Kluwer Academic, London

R. Mesnage, B. Bernay, G.E. S´eralini. (2013). Ethoxylated adjuvants of glyphosate-based herbicides are active principles of human cell toxicity, Toxicology. 313: 122–128, https://doi.org/10.1016/j.tox.2012.09.006

R. Mesnage, C. Benbrook, M.N. Antoniou (2019). Insight into the confusion over surfactant co-formulants in glyphosate-based herbicides, Food Chem. Toxicol., 128: 137–145, https://doi.org/10.1016/j.fct.2019.03.053

Rajna S., Paschapur A.U. and Raghavendra K.V. (2019). Nanopesticides: Its Scope and Utility in Pest Management. Indian Farmer 6(1): 17-21; January-2019

Woods, T.S. (2003). Pesticide Formulations. Encyclopedia of Agrochemicals, New York, pp. 1-11

18

Pesticide Spraying Equipment and Their Operating System

Mood Suguna[1], Majjari Swapna[2], Kammara Meghalatha[3], Nangunoori Maneesha[4] and Kuppireddy Meghana[5]

[1]*Department of Agricultural Entomology, Post Graduate Institute, Dr. Panjabrao Deshmukh Krishi Vidyapeeth, Akola-444104, Maharashtra, India*

[2]*Research Scholar, Department of Agricultural Entomology, Tamil Nadu Agricultural University, Coimbatore-641003, Tamil Nadu, India*

Introduction

Pesticide application plays a major role in effectively stabilizing crop yield by reducing losses to crop pests in agriculture (Cooper and Dobson 2007). Proper technique of application of pesticide and the equipment used for applying pesticide are important to the success of pest control operations. The mains purpose of pesticide application technique is to cover the target species and safety to the non-target organisms and the environment. The complete knowledge of pest is essential for correct time of application. All pesticides are poisonous substances and they can cause harm to all living things. Chemicals are widely used for controlling disease, insects and weeds in the crops (Mayur *et. al* 2023). They are able to save a crop from pest attack only when applied in time. They need to be applied on plants and soil in the form of spray, dust or mist. The important methods of applying pesticides are dusting and spraying. However, spraying is the most common method now because of certain disadvantages associated with dusting like:

i) The loss of chemical due to drift and consequent pollution risk.

ii) Less efficient deposition of dust particles on plant surface resulting in decreased efficiency in pest control.

iii) Tendency of the chemicals in dust formulation to separate itself from the carrier, if its density is widely different from that of the latter, resulting in loss of chemicals.

Dusting will be an useful method of application of chemicals in areas like the vast stretches of unirrigated crop tracts with limited water supply, in hilly terrains and also in case of control of pests occurring in crop just before harvest like ear head bugs and caterpillars. Certain disadvantages with dusting over spraying. They are :

i) Requirement of loss labour for the operation,

ii) Coverage of a larger extent of crop per day than with spraying and

iii) Light weight of dusting application with less risk of corrosion.

Appliances that are used for applying dry dust formulation of pesticides are called dusters. All dusters consist essentially of a hopper which usually contains an agitator, an adjustable orifice or metering mechanism and delivery tubes. A rotary fan or a bellow provides the conveying air. There are two types of dusters.

i) Manually operated dusters

1. Bellows duster
2. Rotary dusters
3. Plunger dusters

ii) Power operated dusters

1. Knapsack dusters
2. High pressure trolley/ Barrow mounted
3. Tractor mounted/trailed duster

Spraying and sprayers: Sprayers are essential tools for the application of pesticides and herbicides in modern agriculture (Aktar *et al.,* 2009). The general principles involved in the spray fluid may be a solution, an emulsion or a suspension of the toxicant. Its liquid phase is usually water or loss frequently, a light oil. However, water is not a good carrier for pesticide because of the hydrophilic nature of the insect cuticle, insolubility of same and breakdown by hydrolysis of certain other insecticides in water and of its evaporation from small droplets thus making the particles light and hence drift. To obtain an effective control of the pest, the toxicant has to be well distributed and to meet this requirement the spray fluid is broken down to fine droplets.in different types of sprays the droplets vary from 30-400 microns in size. The also varies greatly at different distance from the point of emission from the sprayer. Spraying is classified on the basis of the droplet size of the spray as

On the basis of volume of the spray fluid per unit area, spraying is classified as

1. high volume spraying (or) full cover spraying: here droplet size is larger. Spray fluid requirement is 500-1000 lit/ha in case of field crops. For orchard crops (1500-2000).
2. low volume spraying: the low volume sprays are 8-25 times more concentrated than high volume spraying. Droplet size is 70-150μ.

3. ultralow volume spray: air craft or ultralow volume sprayer with special nozzles require 0.5-5.6 lit/ha with droplet size of 20-70 μ.
4. Aerial spraying: Aerial air craft has been employed for application of agricultural and public health pesticides. It is used for spraying, dusting and application of baits. However, spray formulation is more suitable than dusts because of wind speed should not be more than 5 Kmph.it has to be done at low height and in the early hours of the days to ensure uniform deposition of dust particles.

Parts of a typical sprayer

A sprayer consists of important parts are

i) Tank: a built – in tank should have been made of a non- corrosive metal or thick polyethylene and its capacity may range from less than one litres as in a pneumatic hand sprayer to 10 to 25 litres in a knapsack sprayer or mist blower and up to 200 litres in large sprayers.
ii) Pump: it is necessary for location energy required for atomization of spray fluid.

Types of pumps

i) Air pump (pneumatic): The force created by pump acts over spray fluid and the pump does not act directly over spray fluid. They do not pump liquid directly. This type of air pump is commonly used in compression sprayers.
ii) Positive displacement pump: This type of pump is used in plunger, rotary and centrifugal pump sprayers and in this, the pump takes a definite volume of liquid from inlet and transfer it without any escape to outlet.
iii) Agitator: To disperse the pesticide uniformly specially in dusters and the spraying where wettable powder is used for spraying. if the spray fluid is a solution of the toxicant it does not settle and hence does not require to be agitated.
iv) Filters: Filters are provided between pump and spray lance and within lance. The purpose is to protect the pump from abraxion, and to protect blocking of nozzles.
v) Valves: which control the direction of flow of spray fluid.
vi) Hose: for conduction of the spray fluid from the sprayer to the lance and, in the case of sprayer with no built – in tank, from the container to the sprayer, hose pipe is used. It is attached to sprayer one and end spray lance and within the lance. Plastic and nylon hoses are much in use because they are light and cheap.

vii) Spray lance: the nozzle of a sprayer is usually attached to spray lance which has 120^{0}C bend to be useful for spraying under surface of leaf.

viii) Spray cut off – devices: To shut off the flow of liquid usually there is a device in the lance.it may be a stop – cock or a trigger mechanism with provision for adjusting the release of fluid either as a cone or as a jet.

ix) Spray boom: spray burn carrying more than one nozzle is known as spray booms.

x) Nozzle: Nozzle is an important component in a sprayer.it breaks the liquid into droplets and spray outside on the surface. Nozzles performs four types of function (Bayat and Yarpuz-Bozdogan 2005).

- Atomizes liquid into droplet.
- liquid into droplet.
- Disperses the droplets in a specific pattern.
- Meters liquid at a certain flow rate.
- Provides hydraulic momentum.

Selecting a spray nozzle

- The Correct selection and use of spray nozzle is the most important part of pesticide application.
- The nozzle determines the amount of spray that is generated over a given area, the uniformity of the spray produced, the coverage obtained and the amount of drift that occurs.
- Nozzle tips are usually available in brass, stainless steel, and engineering plastic. Steel tips are most resistant to corrosion and abrasion.
- Hollow cone/ hydraulic nozzles: is used for spraying insecticides and fungicides to field crops where complete coverage of the leaf surface is important. At pressures of 40 -8 psi hollow cone nozzles give excellent spray coverage to the undersides of reduces penetration.
- Flat fan nozzles or fish tail nozzles: used for spraying herbicides. the best operation pressure for flat fan nozzle is 15 -30 psi.

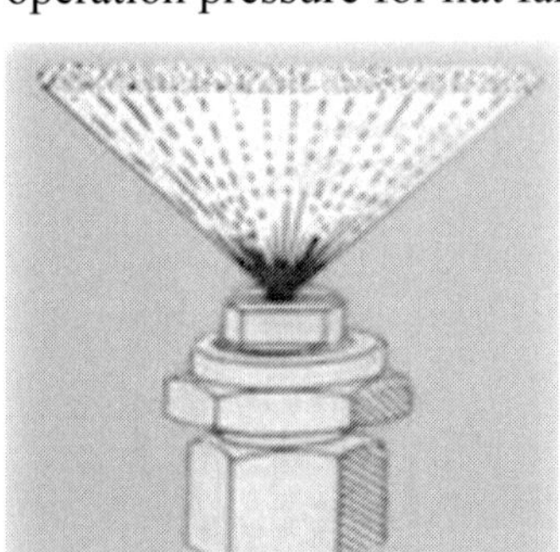

- Flood jet nozzles: Generally, the spray generated by the flood jet is not as uniform as the flat fan type. These are ideal for high application rates and speeds, because they produce a wide – angle, flat fan pattern. These nozzles are also called as deflector nozzle or impact nozzle type. Nozzles are used for herbicide spraying and low pressure 15– 25 psi.

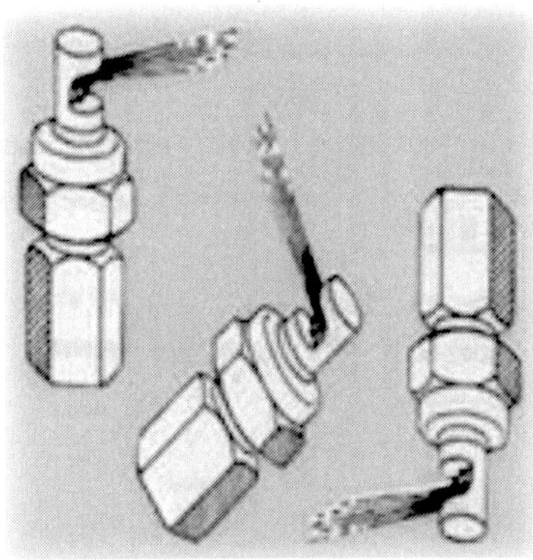

Components of a nozzle

- Nozzle body: it is the main part which enclose all other components of a nozzle.
- Swirl plate: it is metal disc with two tangential holes which imparts a swirl or rotation to the liquid passing through it.
- Nozzle disc: this component breaks the fluid in to fluid into fine droplet. It is a flat disc with an orifice at the centre. When the spray solution reaches the disc from the swirl plate the disc builds up further pressure on the fluid and when the fluid passes out of the orifice it breaks in to fine droplets.
- Strainer: a small circular plastic ring with nylon wire mesh to filter any dust particle coming the chemical solution.
- Spacer: there are two number of runner/plastic ring placed in between nozzle plate and swirl plate and between swirl plate and strainer for effective travel of the solution (nuyttens *et al.,* 2007).

Types of sprayers: the spraying machines are manually operated or power operated. based on energy employed to atomise and eject the spray fluid (Pankaj *et al.*, 2018).

Sprayers are classified into 4 types

- Hydraulic energy sprayer
- Gaseous energy sprayer
- Centrifugal energy sprayer

- Kinetic energy sprayer. In sprayer working with hydraulic pressure, pressure is developed by the direct action on the spray fluid. This pressure force the liquid through the nozzle.

i) Manually operated pneumatic sprayers

1. Hand sprayer (kitchen sprayer): the container for the spray fluid also acts as the pressure chamber. An air pump attached to the chamber project inside. The inner end of the discharge pipe runs down to the bottom of the container and its outlet terminates in a nozzle. It is commonly used for spraying kitchen gardens, in glass houses and indoor against household insects.

2. Knapsack sprayers: they are adopted for spraying large quantities of liquids, otherwise they are similar to pneumatic hand sprayers in the working principle.A typical pneumatic knapsack sprayer comprises a tank for holding the sprays liquid as ‘well as compressed air, a vertical air pump with a handle, filling hole with a strainer, spray lance with nozzle and release and shut off device (Sinha *et al.*, 2018).

ii) Manually operated mist blower: The common air atomizer, often known as a flit gun, is a compact, convenient device. It works on the principle of air blast breaking up the spray droplets and carrying them to the target. It consists of a simple compression cylinder made of tin, with an air pump. These appliances are useful for spot treatment in household spraying against bedbugs, mosquitos and house flies and for treating individual plants.

iii) Manually operated hydraulic sprayers

1. **Hand syringe or garden syringe:** it is a single acting pump working on the principle of a bicycle pump. it consists of a cylinder and a plunger or piston. spray fluid has to be contained in a separate tank. The spray is made of large droplets and is just like drenching. It is useful for small scale spraying in kitchen garden and pot plants.
2. **Bucket sprayer or stirrup pump:** The bucket sprayer is designed to pump the spray fluid directly from, the open container, usually a bucket. The hydraulic pump will be put inside the bucked and held properly with the help of foot rest. As the plunger is pulled up, the fluid enters through the suction ball valve assembly and when the plunger is pressed down, the suction valve closes and the fluid enters the pressure chamber through a ball valve assembly. As the plunger is continuously worked, pressure is built in the pressure chamber and the delivery. A pressure of 4 kg / cm2 is developed in most of the models.
3. **Knapsack sprayers:** the sprayer is used for spraying low crops, vegetable, nursery, shrubs and trees up to 2.5 m high. The capacity of the contains is 15 litres. The pump has to be operated continuously for maintenance of necessary pressure (3-5 kg/sq. cm).
4. **Rocker sprayer:** rocking sprayer has a pump assembly, fixed on a wooden platform with an operating lever, a valve assembly with two ball valves, a

pressure chamber, suction hose with strainer, and delivery hose with spray lance. The operation is continued till the entire suction pipe, ball valve assembly, delivery hose and a portion of pressure vessel is fitted with spray fluid and the pump operator finds it difficult to push the piston forward, due to the downward pressure developed by the entrapped compressed air in the pressure vessel. Thereafter, the trigger cut off valve will be opened to allow the spray fluid to rush through the nozzle and get atomized. Usually 14 to 18 kg/cm^2 pressure can be built in the pressure chamber and hence can be conveniently used for free spray.

5. **Foot sprayer:** This is modified version of rocker sprayer. The pump is fixed in a vertical position with necessary braces. The plunger moves up and down when operated by the pedal. A ball valve is provided in the plunger assembly itself to allow the fluid to cross the plunger and getting pressurized in the pressure vessel. During the upward motion of the piston fluid is sucked in and pressurized into the pressure vessel and during downward movement, the sucked fluid crosses the plungers and enter the pump. The pressure developed is about 17-21 kg/cm^2.

Power operated sprayers

1. **Electrodyne sprayer (electrostatic sprays):** the functioning of the electrodyne sprayer is based on a system which atomizes and propels charged droplets of chemical sprays by electrical forces set up between a nozzle with a positive high voltage charge. Since the charged droplets are readily deposited on a plant, the total deposit on the foliage can be bottle and fed by gravity to the nozzle, where it picks up charged droplets (positive charges) generated by batteries. The liquid leaves the nozzle in a number of uniform ligaments which break up into electrically charged droplets. The positively charged droplets move along curved electrical field line towards and around the plants, covering the visible as well as hidden surfaces. The size of the droplets ranges from 20 and 200μ. a special feature of the type of sprayer is the formulation used in it come in a special bottle fitted with particular nozzle required.it has nozzle (prepacked pesticide + inbuilt nozzle). the insecticide used is cypermethrin.

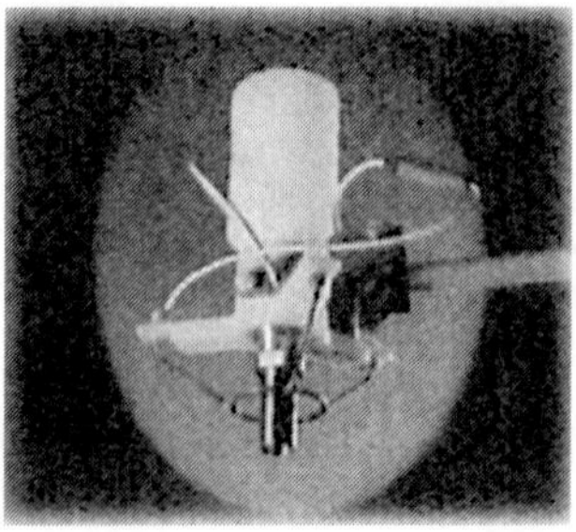

2. **Controlled droplet application sprayer (CDA):** these sprayers apply the correct size and uniform droplets on a given target area so that optimum use is made of spray volume and dosage. It is a logical extension of the ULV concept. The most promising method of controlling droplet size by varying their rotation speed. Uniform droplet density can be obtained by using as little as 500mL/ha of the formulation with a droplet size of 4μ or 1-8 L/ha droplets. The advantages of CDA are it improves spraying efficiency, eliminated waste and protect environment.

Maintenance and Cleaning of Pesticide Appliance

- Avoid bending of the hoses to prevent the pressure reduction and damage to them (Yenisoy-Karakas S 2006).
- Read the instruction manual before using the appliance
- Avoid over heating of the engine
- Use the recommended fuel- oil mixture for running the engine
- Use of recommended lubricant

For dusters

- Blow out all the residual dust at the end of the operation.
- Gear box and gear teeth should be greased occasionally.

For sprayers

Pump out the residual liquid after each operation and then operate the pump with clean water to rinse it and pump out the water again.

- Lubricate the cup leather washer to keep it soft because a dry and wrinkled washer would not produce any suction force.
- Never plunge the piston rashly; this may damage both the cup leather washer and the piston
- Machine should be never be blown with mouth nor cleaned near the source of drinking water
- Operator should never smoke or eat during operation and cleaning of an appliance.
- Cleaning pesticide application equipment is necessary and worthwhile. The risk from not cleaning your spray equipment is reason enough to take the time to do it right. Even a trace of leftover chemical sprayed on a different crop can cause severe damage.
- After each application it is crucial that the sprayer is cleaned to prevent residues from inside the tank contamination the next pesticide used and causing phytotoxicity on the crop.

- Phytotoxicity: the application of pesticides or insecticides on plants is intended to control the pest without causing adverse or harmful effects to plants. It is common to see some adverse insecticides in field which is called phytotoxicity.

There are two types

1. Permanent phytotoxicity: leads to the death of the effected plant part or whole plant.
2. temporary phytotoxicity: which allows the plant to recover after showing phytotoxicity.

Symptoms of phytotoxicity

- Chlorosis or yellow of leaves
- Bronzing of leaves
- Necrosis of complete plant or plant part
- Premature falling of leaves
- Mottle leaves
- Poor germination of seeds
- Scorching

Conclusion

Crop protection practice are most important activities during crop production. Application of pesticides has been increasing in the recent past with advances in technology and changes in farming practices. This chapter will be helpful for the users in selection of suitable sprayer for precise spray as per the need of application. This will not only reduce the cost by application of correct quantity of chemical in appropriate manner but also helpful in the environment by reducing the soil, water and air pollution.

References

Aktar, W., Sengupta, D. and Chowdhury, A. (2009). Impact of pesticides use in agriculture: their benefits and hazards. Interdisciplinary Toxicology, 2(1): 1

Bayat, A. and Yarpuz-Bozdogan, N. (2005). An air – assisted spinning disc nozzle and its performance on spray deposition and reduction of drift potential. Crop protection. 24: 951-960

BETE flat fan nozzle catalogue, www. Bete.com

Cooper, J. and Dobson, H. (2007). The benefits of pesticides to mankind and the environment. Crop Protection, 26: 1337-1348

Mayur, V.J., Yadav, R. and Pravina, S.A. (2023). A comprehensive review of various types of sprayers used in modern agriculture. the Pharma Innovation Journal 2023; 12(4): 143-149

Nuyttens, D., Baetens, K., De Schampheleire, M. and Sonck, B. (2007). Effect of nozzle type, size and pressure on spray droplet characteristic. Bio system engineering, 97: 333-345

Pankaj, B.G. and Shashidhar, S.K. (2018). A Review on pesticides sprayer technology approach in ergonomics, economics and ecologic in agriculture field. International Journal of Engineering Science Invention.7(9): 1-05

Sinha, J.P., Singh, J.K., Kumar, A. and Agarwal, K.N. (2018). Development of solar powered knapsack sprayer. Indian Journal of Agricultural Sciences. 88(4): 590-595

Vasantharaj David, B. and Ramamurthy, V.V. (2017). Elements of Economic Entomology

Yenisoy-Karakas S. (2006). Validation and uncertainty assessment of rapid extraction and clean-up methods for the determination of 16 organochlorine pesticide residues in vegetables. Anal. Chicia Acta, 571: 298-307

19

Plant Disease Vector and Their Management

Saurabh Singh[1], Brajrajsharan Tiwari[2], Vivek Singh[3] and Himanshu Kumar Gupta[4]*

[1,2]*Department of Entomology College of Agriculture, Banda University of Agriculture and Technology, Banda-210001, Uttar Pradesh, India*

[3]*Department of Plant Pathology College of Agriculture, Banda University of Agriculture and Technology, Banda-210001, Uttar Pradesh, India*

[4]*Department of Plant Pathology, Veer Bahadur Singh Purvanchal University Jaunpur-222003, Uttar Pradesh, India*

History

- The fossil evidence also indicates the presence of plant diseases 250 million years ago.
- The diseases caused by fungi and nematodes were discovered by Monocern in 1735.
- Johan Needham (1743) identified wheat gall caused by a nematode.
- Adolf Mayer (1886), demonstrated the transmission of Tobacco mosaic virus (TMV).
- EF Smith (1890) described bacterial wilt in cucurbits.
- The thrips (Thysanoptera) as vectors of plant viruses were identified the scientific studies made by Samuel and his team.
- Karl Maramorosch (1952) demonstrated the multiplication of Aster yellows virus (AYV) in leafhopper vector.
- The mite Aceriatulipae has been identified as a vector of Wheat streak mosaic virus by JT Slykhuis in 1955.
- The term "viroid" (viruses without protein coat) was coined by TO Diener during 1971.
- Single stranded Tobacco mosaic virus (TMV) was identified in 1982.

Mode of transmission the transmission of plant pathogens takes place through two different mechanisms under natural conditions (vertical transmission and horizontal transmission).

Horizontal transmission Horizontal transmission is carried by external agents such as insect vectors, human activities, pruning, agriculture tools, and other direct or external contamination. In this pathway of transmission, viruses are directly or indirectly transmitted among individuals of the same generation. In direct horizontal transmission, plant viruses are transmitted by a direct route such as airborne infection and food- borne infection, whereas in indirect transmission, plant viruses are transmitted through intermediate biological vectors such as aphids, whiteflies, and leafhoppers, which acquire and transmit virus from diseased plant to healthy plants.

Vertical transmission

Vertical transmission occurs when a virus's goes in to the progeny from the infected parent's plant either through asexual propagation such as cuttings, grafting, and budding etc. or in sexual reproduction via infected seeds. In this pathway, viruses are passed vertically from mother to offspring via egg, either on the surface of the egg, called "transovum transmission," or within the egg, called "transovarian transmission" (Clayton and Tompkins, 1994; Ewald, 1994).

Insect transmission

- 400 spp. have been reported to transmit plant viruses.
- 94% of these belong to Phylum Arthropod in this 90 % are insects
- 70% of insects belong to Homoptera in which Aphididae are the most important group.
- Beetles, thrips, mites etc. are also important vectors.

Insect vector for plant viruses Most plant viruses depend in insect vectors for their survival, transmission and spread. Specific virus. Insect vectors interactions have evolved, in a mutualistic symbiosis at the detriment of the plant. The principal families of insect vectors which cause the most damage to agricultural crops through the spread of plant virus are in the order Hemiptera, and include the aphids, leafhoppers, delphacid planthoppers and whiteflies etc.

Mechanism of Transmission

1. **Non-persistent or stylet borne transmission:** The virus binds in the stylet during feeding and is released when the insect secrets saliva on a new feeding place, mostly aphids.
2. **Semi-persistent transmission:** Viruses are retained in the foreguts or salivary glands. Such as whitefly.
3. **Circulative, non-propagative:** The virus circulates up through the food canal and cross the midgut and hindgut reaching the haemocoel. The virus

particles cross accessory salivary glandsand return to the plant via the saliva canal, Some spp. of aphids and leaf hoppers.

4. **Persistent propagative transmission:** The virus circulates in the host as described above, but actually infects insect cells and replicates in the vector, mostly leaf hoppers.

Table 1. Transmission characteristic and timing of plant viruses transmitted by Hemipteran insects

Biological characteristic	Non-persistent	Semi- persistent	Persistent-circulative	Persistent
AAP and IAP	Seconds, minutes	Minutes, hours	Hours, days	Hours, days
Latent period	None	None	Hours, days	Days, weeks
Retention time in vector	Minutes,	Lost after moulting	Hours, lost after moulting	Days, weeks
Presence in vector's hemolymph	No	No	Yes	Yes
Multiplication	in vector	No	No	No
Transovarial transmission	No	No	No	Often

***AAP** – Acquisition access period * **IAP**- Inoculation access periods

Table 2. Important insect vector

Sl. No	Insect species	Disease	Plant	Mode of transmission
		Aphids		
1.	*Myzus persicae*	Cucumber mosaic v.	Cucumber	Non-persistent
		Beet yellow mosaic v.	Beet	
		Barley yellow dwarf v.	Barley	
		Potato leaf rollv.	Potato	
2.	*Pentalonia nigronervosa*	T.M.V.	Tobacco	
		Bunchy top	Banana	
		Kokkan disease	Banana	
3.	*Aphis craccivora*	Papaya mosaic virus	Papaya	
		Rosette virus	Groundnut	
		Mosaicvirus	Cowpea	
4.	*Toxoptera aurantii*	Tristeza	Citrus	

		Whitefly		
5.	*Bemisia tabaci*	Y.M.V.	Okra	Semi-persistent
		Yellow mosaic	blackgram	
		Leafcurl	Tobacco, Tomato, cotton, papaya	
		Yellow mosaic	Soybean	
		GLH		
6.	*Circulifer tenellus*	Beetcurlytopv.	beet	Persistent circulative
7.	*Nephotettix virescens*	Ricetungro	Rice	Persistent propagative
8.		Yellowdwarf (MLO)	Rice	
9.	*Orosius albicinctus*	Phyllody (MLO)	Sesame	
9.	*Cestiusphycitis*	Littleleaf (MLO)	Brinjal	
		BPH		
11.	*Nilaparvata lugens*	Grassystunt	Rice	Persistent circulative
		Grassy stunt (MLO)	Sugarcane	
		Thrips		
12.	*Frankliniella schultzei*	Bud necrosis virus	Groundnut	Persistent Circulative
13.	*Frankliniella occidentalis*	Spottedwilt virus	Tomato	
14.	*Scirtothrips dorsalis*	Yellow spot virus	Groundnut	
		Chilli leaf curl	Chilli	
15.	*Thrips palmi*	Bud necrosis virus	Watermelon	
		Plant bug		
16.	*Anasa tritis*	Yellow vein (bacteria)	Cucurbits	Semi- persistent
17.	*Peregrinus maidis*	Mosaicvirus	maize	
18.	*Diaphorina citri*	Greening disease	Citrus	
		Leaf minor		
19.	*Phyllocnistis citrella*	Canker (bacteria)	Citrus	Non- persistent
		Beetle		
20.	*Aulacophora foveicollis*	Mosaicvirus	Bottle guard	Non- persistent
21.	*Sesselia pusilla*	Yellow mosaic	Rice	
		Mite		
22.	*Aceria tulipae*	Streak mosai cv.	Wheat	Non- persistent
23.	*Aceria cajani*	Sterility mosaic	Pigeon pea	
25.	*Tetranychus urticae*	Potato virus Y	Potato	

Management

Destruction of volunteer plants and alternate hosts: The volunteer plants and several weed hosts are known to serve as a reservoir for viruses and these plant species become primary sources of infection. Russian thistle is an overwintering host, both for virus, Sugar beet curly top virus, and its vector, sugar beet leafhopper *Circulifer tenellus*. There are many viruses vectored by aphids, whiteflies and leafhoppers that have reservoir hosts suc has *Physalis* sp. (Rugose mosaic virus in potato), *Hibiscus tetraphyllus* (Yellow vein mosaic virus in okra) and *Datura* spp./ *Cantharanthus* sp. respectively. The uprooting of cotton sticks is a practice used to eliminate the sprouting problems of cotton in many areas in the world. In Sudan, there is a regulation pertaining to the destruction of reservoir hosts of Cotton leaf curl virus with a view to eliminate the source of inoculum. There moval of alternate reservoir hosts of Wheat streak mosaic virus is also being practiced to tackle this problem of virus disease in the wheat crop at many places.

Peach yellows and Peach rosette viruses are being kept under check with the destruction of alternate hosts. In South India Santhyamanglem, the practice of destruction of diseased trees of sandal wood is being followed in order to reduce the further spread of Spike disease of sandal wood. The practice of destruction of weeds from banana orchards is followed to get rid of Banana bunchy tops virus in South India.

Rouging of diseased plants/ parts of the plant: It is common practice to delay the spread of plant viruses.

To follow this practice, it is advisable to uses lightly higher seed rate at sowing. This practice is common in cotton in North India and it is being followed on a large sale to mitigate the problem of leaf curl disease in cotton. The rouging of diseased plants in cassava (Cassava mosaic virus), green gram (Yellow mosaic virus), citrus (*Citrus tristeza*), cardamom (Katte disease) and banana (Bunchytops of papaya virus) is being practiced regularly to mitigate the disease severity. The surgical approach is sometimes helpful in reducing the disease. There moval of bunchy tops of papaya and decapitating of trees prevented the *Empoasca papayae* vector from spreading the bunchy top virus. In addition to papaya virus, surgery also proved useful in preventing the Witches broom in Willow.

Modification of harvesting dates: A specific technique has been developed in India that can be emulated in other situations throughout the globe to tackle aphid vectors of plant viruses. In earlier production of virus-free seed of potato, the potato seed production programme in India was mainly confined to the hill regions because of absence of aphid vector population during the period of growth of potato. With scientific progress, the arrival of aphids was analyzed in the plains of the state of Punjab. Now the seed production programme is taken up in plains of Punjab through seed plot technique instead of in the hills. In this technique, the potato seed production is done during the period of absence of aerial vectors by advancing the planting of potato and cutting of halms in relation to the appearance of aphid vectors in Punjab plains. The crop sowing is done in the last week of September and halms are cut before 20th December (when the aphid vector begins to appear). Once the halms are cut, the aphid vector is unable to pick up the virus from the stems. The harvesting can therefore be taken up at the convenience of the farmer. The potato seed produced via this technique is free from almost all viruses. Based on the data of aphid arrival in plains for thirty years, the appearance of aphid vector is expected by 20th of December or later.

Planting time/ planting density: A higher planting density in cotton is being recommended to manage the Cotton leaf curl virus-CLCuV in North, India which is a limiting factor in the successful cultivation of cotton. Also, leaf curl resistant hybrid Fateh has been developed to check this malady in cotton. The closer planting of tomato is quite useful in checking the incidence of TbLCuV transmissible by whiteflies. Similarly, Cucumber mosaic virus- CMV in cucurbits (*Aphis gossypii*) and Rosette virus in groundnut (*Aphis craccivora*) transmissible through aphids are managed with dense planting. Tillage-Minimum tillage has been found to be better than conventional tillage in reducing the incidence of Tomato spotted wilt virus as well as population of thrip vector, *Frankliniella occidentalis*. The thrips were unable to detect the crop due to no tillage and the presence of crop debris of the previous crop.

Intercropping/mixedcrops: To cover the risk of adversity of nature, two or more crops are planted and mixed with the main crop. This practice is quite common in minimizing the incidence of non- persistent plant viruses. The cucumber, being the preferred host of *Bemisia tabaci* and non- host of Tomato yellow leaf curl virus, is planted about one month earlier so that whitefly should remain confined to cucumber only where the se whiteflies are controlled through the use of pesticides. The mixed crops (3 rows of maize/ sorghum/pearl millet) around/between the main crop (blackgram/greengram) are another important method to keep the main crop free from yellow mosaic virus diseases. Mixed cropping of carrot and onion in England (Uvah and Coaker, 1984) are instrumental in reducing the population of thrips to the tune of 80% and 20%, respectively.

Use of trap crops

- The use of yellow sticky traps management of aphid and whitefly.
- The reflection of these mulches is instrumental in the killing of whiteflies due to heating effect.
- Use of light traps.

Use of fertilizers: The fertility is known to influence the population of insect vectors. The reduction in the dose from 402 kg/ha of nitrogen in onion crop is instrumental in reducing the population of thrips (Buckland *et al.*, 2013). The application of phosphorus (P2) reduces the population of whitefly *Bemisia tabaci* in cotton as the leaves become thickened and adult whiteflies do not prefer such a heavily fertilized crop (Butter *et al.*, 1992b).

Chemical: Chemicals are applied in order to manage the colonizing vectors in a crop, but rarely the invading population of immigrant vectors from outside the primary crop. The goal of managing the colonizing population of vectors is achieved and the desired control is obtained. Despite the availability of different chemicals, insecticides are not a priority tactic in managing non-persistent plant viruses. The non-persistent and semi-persistent viruses require a short inoculation period. Under these circumstances, the use of chemicals increases the incidence of viruses. The pesticide application enhances movement of vectors and increases the intensity of probing which in turn increases the incidence of stylet-borne viruses. These chemicals also kill the natural enemies of vectors. As a result, the vector population increases beyond the limit and causes extensive damage due to the large population, by way of the spread of viruses. However, the use of insecticides proved better against circulative/propagative viruses. The chemicals used to manage vectors include chemicals of plant origin, repellents, use of baits, mineral oils, behavior manipulating chemicals, antifeedants and insecticides. Of these groups of pesticides, repellents and baits appear to be more promising. However, these have not been exploited to the desired level. Furthermore, insecticide use in combination with other tactics can be still more fruitful.

Conclusion

The plant pathogens are dependent of various agencies for their spread and survival. Among the various plant pathogen transmitting through the vectors, insect vectors are prominent ones. ETL level can't be decided in the insect vector. One individual vector can be transfers many pathogens in plant.The management of insect vectors with the use of various tactics should be used to reduce the severity and to delay appearance of disease.

References

Buckland K., Reeve J.R., Alston D., Nischwitz C. and Drost D. (2013). Effects of nitrogen fertility and crop rotation on onion growth and yield, thrips densities, Iris yellow spot virus and soil properties. *Agriculture, Ecosystems & Environment*, 177: 63-74

Butter N.S., Kular J.S., Singh T.H. and Chahal G.S. (1992b). Management of whitefly, *Bemisia tabaci* Genn in cotton through the soil application of phosphorus. Proceedings of the Vasant Rao Naik Memorial National Seminar on cotton production in India, 5th–6th December 1992, Nagpur, pp. 127-135

Griffiths D.C., Pickett J.A., Smart L.E. and Woodcock C.M. (1989). Use of insect antifeedant against aphid vectors of plant virus disease. *Pesticide Science*, 27: 269-276

Harris K.F., and Maramorosch K. (2013) Vectors of plant pathogens. Academic Press; Oct 2

Jones T.K., and Medina R.F. (2020). Corn stunt disease: an ideal insect–microbial–plant pathosystem for comprehensive studies of vector-borne plant diseases of corn. Plants. Jun 14; 9(6): 747

Uvah III and Coaker T.H. (1984). Effect of mixed cropping on some insect pests of carrots and onion. Entomologia Experimentalis et Applicata, 36: 159-167

20

Damage Caused by Crop Raiding Wild Boar and Their Management

Mukesh Kumar Mishra[1*], Rakesh Pandey[2] and B.K. Singh[2]

[1]Department of Entomology, Banda University of Agriculture and Technology Banda-210001, Uttar Pradesh, India

[2]Department of Entomology, Banda University of Agriculture and Technology Banda-210001, Uttar Pradesh, India

Introduction

In agricultural and horticultural ecosystems greater extent of losses caused by pests includes insect-pests, plant diseases, and weeds and have directly affected the production. Out of these, a group of animals other than the insects also causes the considerable yield losses to agricultural and horticultural crops and commonly known as vertebrate pests such as birds, rodents and mammals. In recent times mammals such as blue bull (Nilgai), wild boars, rodents, porcupine and monkeys has gaining the status of pests and in certain cases a huge damage is being encountered (Rao, *et al.*, 2015). Among mammals, blue bull and wild boar both has become a serious problem in most of the Indian states for farmers in major agricultural and horticultural crops and resulting into immeasurable damage. Of the both, the wild boar (*Sus scrofa* L.) also known as the "wild swine" is one of the utmost broadly spread ungulates and large mammals' pest in the biosphere owing to its great multiplicative speed, adaptableness, and resourceful nourishing (Ballari and Barrios, 2014) and has turned out to be a consistent threat to agriculturalists in case of most important plants causing into massive losses (Muhammad Sarwar, 2019). Destruction of natural habitats by escalating rate of deforestation can be basic reason to abrupt raise in wild boar populations. Deforestation also resulted in the decline the population of natural predators such as Tigers, Panthers, Wild dogs, Wolf, and Jackal etc. thereby indirectly contributing to the phenomenal raise in the wild boar populations. (Khokhar and Rizvi, 1998). Over exploitation of forest resources by the mankind forced wild boars out of their natural habitat and compelled them to depend on cultivated agricultural crops like Rice, Maize, Sorghum, Pulses, Oil seeds and horticultural crops (Fruits and Vegetables).

The damage caused by wild boar is more alarming than their actual feeding in the crop and more apparent in agricultural fields with close proximity adjoining forests areas (Rao *et al.*, 2015).

To develop an effective management strategy for the management of wild boar, the basic needs to know about its origin, distribution, habitat, feeding behaviour and reproduction patterns.

Scientific or Taxonomic Classification

Common name	:	Wild boar, wild swine, wild pig.
Kingdom	:	Animalia
Phylum	:	Chordata
Class	:	Mammalia
Order	:	Artiodactyla
Family	:	Suidae
Genus	:	*Sus*
Species	:	*S. scrofa*
Binomial name	:	*Sus scrofa Linnaeus*

Subspecies: As of 2005,16 subspecies are recognized, which are divided into four regional groupings (Wozencraft, 2005).

Western: *S. s. scrofa, S. s. meridionalis, S. s. algira, S. s. attila, S. s. lybicus and S. s. nigripes.*

Indian: *S. s. davidi and S. s. cristatus.*

Eastern: *S. s. sibiricus, S. s. ussuricus, S. s. leucomystax, S. s. riukiuanus, S. s. taivanus and S. s. moupinensis.*

Indonesian: Represented solely by *S. s. vittatus*, it is characterized by its sparse body hair, lack of underwool, fairly long mane, a broad reddish band extending from the muzzle to the sides of the neck.

Origin and Distribution

Wild pigs were first introduced in California in 1769 by the Spanish missionaries at Mission San Juan Bautista Wild boar is a suid native to much of Eurasia and North Africa, and introduced to the America and Oceania. According to Chen *et al.,* 2007, the wild boars obviously originated in South East Asia during early Pleistocene. Except Antarctica and many oceanic islands this species occurs purely wild or narrowly modified feral form on all continents. It is found throughout the plain and broad-leaved forest regions of the Palaearctic from western Europe to the east while soviet far East, extending southwards as far North Africa. The Mediterranean basin and middle east, through India,

Indo-China, Japan, Taiwan and Greater Sudan Islands of South East Asia. It is common in the whole of India except Jammu and Kashmir, upper Himalayas and desert, Sri Lanka, Burma and Tenasserim, Syria and part of the Malya Peninsula.

Status under Indian Wildlife (Protection) Act, 1972: Wild boars is declared wild under Schedule III and is protected, but their hunting does not attract high penalties or severe punishments.

Habitat and Feeding behaviour

Habitat Regions: Temperate and terrestrial

Terrestrial Biomes: Savanna or grassland forest

Wetlands: Marsh and swamp

Other Habitat Features: Agricultural and riparian

The wild boar *Sus scrofa* is a social species with a strong post weaning association between mothers and daughters lasting several years with some fluctuations throughout the year (Kaminski *et al.,* 2005). They inhabit a diverse array of habitats from boreal taigas to deserts. The wild boar even occupies alpine zones in mountainous regions, that occurs up to 1,900 m (6,200 ft) in the Carpathians, 2,600 m (8,500 ft) in the Caucasus and up to 3,600–4,000 m (11,800–13,100 ft) in the mountains of Central Asia and Kazakhstan (Heptner *et al.,* 1988). The wild boars require a vast range of habitats with heavily brushed areas providing shelter from predators, water for drinking and bathing purposes and irregular snowfall. Being widely distributed in India, wild boars prefer grasslands, scanty bush jungles, forests and agricultural ecosystems. They generally move in groups and the group size varies between 15-35 and home range between males and female.

The wild boar is omnivorous in nature, whose diversity in choice of food is comparable to that of humans. They feed on Rhizomes, roots, tubers and bulbs, nuts, berries and seeds, leaves, bark, twigs and shoots, along with garbage. Beside it they also feed on earthworms, insects, mollusks, fish, rodents, insectivores, bird eggs, lizards, snakes, frogs and carrion in warm periods. Wild boar causes a lot of damages to almost all the agricultural field crops throughout the world. About 400 plants species have been recorded as the wild boar's diet, of these, 40 species are crop plant. The wild boar injury is further upsetting than their real nourishing in the crop (Chauhan, *et al.,*1993). Wild boar able to identify the host plants by their smell detection sensory organs and more pronounced feeding has done in dawn and dusk periods. By using the crops for both food and shelter, the wild boar caused serious injury to the crops. In many of the states, the wild boar (*S. scrofa*) is frequently linked with significant crop losses.

Reproduction in Wild Boar

Wild boars are prolific breeders will breed year-round, but births peak in November to January in most of the areas. The reproductive period in wild boar is seasonal and mostly correlated with availability of food and other climatic factors. Prior mating, the males develop their subcutaneous armour in preparation for confronting rivals and travel a long distance for searching a sounder of sows, eating little on the way. The males sexually mature in 15 to 24 months but they mate with female at five years of age, when reached their full size. However, females have mate at 18 months of age. Females have an estrous cycle of 21 days and generally receptive for 2-3 days. The gestation period for first-time breeders, it lasts 114–130 days, while it lasts 133–140 days in older sow and a sow will give birth to anywhere from 5 to 12 little piglets of 600–1,000 grams body weight in March and May (Heptner *et al.,*1988). She can have 2 litters a year and mostly seen after the rains. The newly borne piglets fades three months after their birth, with adult colouration being attained at eight months. The lactation period of new is about 2.5–3.5 months, the piglets begin displaying adult feeding behaviour after 2–3 weeks. The permanent dentition is completed in 1–2 years. The male wild boars leave their sounder after 8–15 months, while females either remain with their mothers or establish new territories nearby. Wild boars are potentially dangerous because of their size and their tusk but they don't attack unless provoked. The maximum lifespan of boar in the wild is 10–14 years with captivity have lived for 20 years (Baskin and Danell, 2003).

Table 1. Morphological characteristics of Wild boar and Domestic Pigs

Characteristics	Wild Pigs	Domestic Pigs
Hair	Amply covered with coarse, long hair	Sparse, short hair
Ears	Relatively small and erect	Relatively large and floppy
Tail	Straight, covered in hair	Curly, little hair present
Body	Razor-backed, shoulders higher and wider than hindquarters	Wide body, flat back
Tusks	Long and sharp	Relatively short
Head	Longer snout with flat profile	Shorter snout, concave profile
Color	Mostly black, some pied or russet	Usually white, sometimes russet or pink
Young	Dark with horizontal stripes	Same uniform color as parent

(*Source*: Davis, C.A. in Agriculture and Natural Resources, July 2015)

Crop Raiding Behaviour of Wild Boar

Wild boar may destroy to agricultural and horticultural crops by crushing or eating of plants. They harm to crops by injury to plant roots, and form holes and grooves that harm farm equipment and endanger operations. Wild boar injury is much higher in the fields, that are in neighboring vicinity of forests ranges. They harshly influence trees and timber resources in many ways like disturbance of planted seedlings, girdling and rubbing of matured trees, chewing and rooting of sideway roots of fully-fledged trees, and tusking or scent marking with the tusk glands that injuries to tree bark (Sarwar, 2019). According to Chauhan, 2011; about 927 human casualties were recorded during 1990-2010 in 11 states by wild pigs with 4.2% death and 95.8% injury cases. Among the causality's cases maximum (81.2%) were recorded as male followed by 18.8% in female causalities. Among all the observed cases, the maximum cases were occurred under forest areas (77.9%) followed by 18.3% cases under cropped fields and 3.8% cases in the vicinity of village.

Integrated Management of Wild Boar

Wild boar can be managed through small-scale exclusion, trapping, and/or shooting. While recreational hunting is often a preferred method, it is not effective at controlling wild boar population growth. In Indian conditions, wild boars have been listed in Schedule III of Wild Life Protection Act 1972, hence killing methods are not applicable in its management because of any method resulting into death of the animal is directly punishable. Under such situation use of strategic and logical integrated management practices are effectively minizine the crop injury caused by wild boar and also farmer – animal convict in agricultural ecosystem.

1. Physical barrier

This particular character is exploited for developing certain types of physical barriers to save the crops from this noxious animal.

Wire fencing

- Fencing the field with 3 rows (1st row fitted 12 inch above the ground) of barbed wire forming a physical barrier to many of the stray animals as well wild boars entering into the cropped area.
- Fenced the field with barbed wire on farm periphery with installing cement poles at 10' (3.05 m) interval and tie 7 parallel rows of barbed wire one above other and 2 rows diagonally crossing each other at center between two adjoining poles to restrict boar (*Sus scrofa*) entering into crop field. The parallel rows of barbed wires, starting from 6" (15.24 cm) above ground, lower 4 rows 8" (20.32 cm) apart and upper 3 rows 12" (30.48 cm) apart.
- Fencing of field by fixing Razor blades with iron wire at regular interval around the cropped area as border by forming circular rings may cause serious injury to wild boar, when they try to enter the field.

- Chain link meshes of 3 feet height can be fixed around the crop by maintaining a distance of 1ft away from the cropped area.

Plastic fencing

Polypropylene fencing is a less expensive and easier to install and repair than the other type of fencings and it is widely acceptable and meet various regulations. It is a short-term (up to 10 years) fencing method and comparatively less effective in wild animals' endemic areas.

Solar fencing

Solar fencing is a modernized and unconventional fencing method and it is one of the most effective and efficient options of providing security. It is not only doing the guarantee property protection, but also uses renewable of solar energy for its functioning. The solar fencing works like an electric fence with 12-volt (DC) electricity, which delivers a brief yet fierce shock when any animal including wild boar or human come in contact with the fence. The shock of solar fencing is provide a deterrent effect, that ensuring to no loss of the life.

Electric fences

The electric fencing is acting as a barrier against the wild animals, cattle and human to entering in the fenced area. It generates an electric shock to deter any animals or people from crossing a boundary. The voltage of the shock may have effects ranging from discomfort up to the death. The electric fencing is also useful for controlling the movements of wild boar from raiding crops. This is a long-term fencing and very effective against the wild animals to protection protect the crops. The costs of electric fencing is variable and depending on specific type and size of an area.

Trenching

Wild Boar starts work with hand trenching and pitting in the field. If the farmers can make the trenches of (depth of 8ft with bottom width 3.25ft and 7ft upper width) around the cropped area then the wild boar will fall into the trench, when they try to cross the trench and the forest department personnel will easily rescue it (Kumaran, 2015).

2. Bio fencing

Bio-fencings is the lines of trees or shrubs planted around the field boundaries that provide protection to crop against wild animals or cattle. It also acts as windbreaks, enrich the soil, bee forage, shading to crop and control the dust. These are less expensive and more useful than the wooden, barbed wire or stone masonry fencing. It is ecofriendly in nature giving additional income to the farmer through their produce. Various species have been tested to discover their suitability for use as bio-fencing plants ex. thorny species have been widely used.

- Use of four rows of Safflower as border crop around ground nut found to be most promising in preventing the damage by wild boar. At the first instance wild boar will fail to locate the crop or if it is locating, the thorns of safflower cause injury when they trying to enter into the crop specially in groundnut.
- Planting of 4-5 rows of castor crop around the maize field with close spacing helps the protection against wild boar. Because of the wild boars being capable of identifying maize crop only through smell cannot do sowing to the strong odour emitted by the castor successfully masking the odour emitted by the maize crop. Castor can be used as border crop also in pulses and oil seeds.
- Planting of thorny bushes and xerophytes viz., *Cacti* sp., *Euphorbia caducifolia, E. meriifolia, Opuntia elatior, O. dillenii, Ziziphus mauritiana* and *Agave americana, Caesalpini acriatata* on around the crop helps to protect the crop from wild boars due to their thorny nature.
- Planting of Karanda around the crop field planting of karanda (*Carrissa carandus*) around the crop as bio-fence does prevent effectively the entry of wild boars into the cropped area owing the thorny nature.
- The grey wolf is the ferocious predator of wild boar throughout its range and a single wolf can able to kill upto 50 to 80 wild boars of differing ages group in a single year. Human are the main predator of wild boars. Wild boars also be destructive to farmland and natural ecosystems causing humans to implement removal process. Mostly the young wild boars are targeted by predatory coyotes and bobcats, while juveniles and adults may fall prey to American black bears and cougars like larger predators (Chapman and Trani, 2007); Dholes, leopards and tigers also prey on boars.

3. Trappings

The maximum effectual means for eliminating wild boars is trapping for the reason that it is a constant action needing far a reduced amount of time and work than other approaches, for instance, hunting. A successful boar trapping pivots upon more than a few following important machineries We captured wild boar in big cage traps of 5 x 2 x 2 m by video-observation and manual (electromagnetic) release.

4. Using of Dogs

Wild boars hunting through dogs has grown into a very widespread method of sport hunting. This technique of hunting is controversial, it helps as a prized device while demanding to eliminate boars in areas where extreme gunfire and catching stress have compelled boars to develop further nightly in their activity pattern.

5. Local methods

- Spraying of dung solution of local pigs around the crop by 1ft width at seven days interval confuse wild boars with a false assumption of entering into the territory of local pigs or other animals, there by their movement will be prevented to avoid territorial conflict.
- Use of human hair in the movement route of the wild boar gets sucked through nostrils causing severe respiratory irritation and disturbed to loses its track by making distress calls, which will ward off others wild pigs are entering into the cropped area.
- Arrangement of colourful used sarees around the crop can make wild boars to falsely assume human presence in the area thereby they avoid to enter into cropped areas.
- Burning of dried dung cakes with local pig dung helps in sensitizing wild boar about the un existing presence of local pigs and the wild boars avoid to movement into such areas.
- Using the trained dogs on a community basis in endemic areas scared away the approaching wild boars to attacks. This is also effective and sustainable.
- Using the fire crackers, sounds through local drums or empty tins, making bonfires and shouting on community basis help to scared away the wild boar's form cropped area.

6. Bioacoustics

The bio-acoustic technique uses only the sounds of predators, distress and alarm calls of target and closely related species of the target animals. The calls are broadcast in a field by suing an electronic platform with sound drives. Wild Boar reppeler is an impressive and state of the art device that uses different type of sounds at regular intervals to keep wild boars away from the cropped area. This device consists of 1 or 2 loudspeakers, an electric control module with cable, battery terminal clips for power supply and a fixation-plate. That produces fixed volume of 110 dB, and 4-5 acres area at 42 dB ambient noise level and upto 19 acres at 37 dB ambient noise level. This device is perfect for the farmers for protecting agricultural area. A newly designed Duet sonic technology allows the Ultra-Duet Pest repeller, that emitting ultrasonic waves by the help of two different sound systems simultaneously and able to deter and keep wild boar and other stray animals away from house, garden and cropped area.

6. Chemical methods for wild boar management

Wild boar management by killing through chemical is easy practice but cannot be practiced due to listing of wild boars in schedule III of Wild Life Protection Act 1972. The following are the some of such chemical methods that are worked out to manage the wild boars truly under the limits of Wild Life Protection Act.

- Arrange the granular insecticides with sand around the crop at 60-100cm height and 3 mts by tied in perforated polythene bags helps to prevent the crops from wild boar attack due to slow release of the smell by wind currents. Strong smell of insecticide may also help in the masking of natural smell of the crop, there by confusing the wild boars in locating the crop.
- Wild boar using smell of the crop as criteria for identification, hence the spraying of egg solution (20 ml/l of water) either on the border row of the crop or on the wet soil around the crop is capable of successfully making the natural odour of the crop and thereby reducing the wild boar damage. Under haviour crop damage again repeat it at 10 days interval.
- Arranged the 3 rows of niwar around the crop after socked in kerosene oil for 2hrs at 1feet row to row distance with the help of wooden poles. The wild boar could not identify the crop due to dominating smell of the kerosene and it is generally effective for 10-15 days.
- Arrange the 3 rows of coconut ropes soaked in the mixture of sulphur + pig oil around the crop at 1 feet row to row distance and 10 days interval. It generates the typical smell there by repelling wild boars, thus they do not enter into the cropped area. in between.
- Use of poison baits with non-lethal doses of monofloro acetate (0.025%) or warfarin (0.025%) cause stomach irritation in wild boars after consumption. However, pre-baiting with 100 gms of wheat flour mixed in onion and garlic paste should be done continuously for a about 3-4 days by keeping 10-15 baits on field bunds, which makes the wild boar to feed.

References

Ballari, S.A. and Barrios-G, M.N. (2014). A review of wild boar Sus scrofa diet and factors affecting food selection in native and introduced ranges. Mamm. Rev., 44: 124-134

Baskin, L., Danell, K. (2003). Ecology of Ungulates: A Handbook of Species in North, Central, and South America, Eastern Europe and Northern and Central Asia. Springer Science & Business Media. pp. 15–38. ISBN 3-540-43804-1

Chapman, B., M. Trani (2007). Feral Pig (Sus scrofa). Pp. 540-544 in M Trani (Griep), W Ford, B Chapman, eds. The Land Manager's Guide to Mammals of the South. Durham, NC: The Nature Conservancy and the US Forest Service, Southern Region

Chauhan, N.P.S. and Rajpurohit, K.S. (1993). Survey of Animal Damage Problem in and around protected areas and managed forests: Phase- I Madhaya Pradesh, Bihar and Orissa - A Report, Wildlife Institute of India, Dehradun

Chauhan, N.P.S. (2011). Human casualties and agricultural crop raiding by wild pigs and mitigation strategies in India DO - 10.5073/jka.2011.432.105

Chen K., Baxter T., Muir W.M., Groenen M.A. and Schook L.B. (2007). Genetic Resources, Genome Mapping and Evolutionary Genomics of the Pig (Sus scrofa). Int J Biol Sci. 3(3): 153-165

Coblentz, B.E. and D.W. Baber (1987). Biology and control of feral pigs on Isla Santiago, Galapagos, Equador. Journal of Applied Ecology. 24: 403-418

Heptner V.G., Nasimovich A.A., Bannikov A.G., Hoffman R.S. (1988). Mlekopitajuščie Sovetskogo Soiuza. Moskva: Vyssaia Škola [Mammals of the Soviet Union]. I. Washington, D.C.: Smithsonian Institution Libraries and National Science Foundation. pp. 19-82

Kaminski G., Brandt S., Baubet E. and Baudoin C (2005). Life-history patterns in female wild boars (Sus scrofa): mother-daughter postweaning associations. Can J Zool 83: 474–480

Khokhar A R. and Rizvi, S.W.A. (1998). Productivity enhancement of rice crop yield through prevention of losses due to wild boars in Pakistan

Kumaran, S. (Nov 21, 2015). 90 km trench dug in Erode to prevent boar-human conflict http://timesofindia.indiatimes.com/articleshow/49868005.cms?utm_source= contentofinterest & utm_medium=text&utm_campaign=cppst

Muhammad, S. (2019). Raiding of agricultural crops and forests by wild boar (Sus scrofa L.) and its mitigation tricks. Journal of Scientific Agriculture 3: 1-5

Rao, V.V., Naresh, B., Reddy, R.V., Sudhakar, C., Venkatachalam, P. and Ramarao, D. (2015). Traditional management methods used to minimize wild boar (Sus scrofa) damage in different agricultural crops at Telangana state, India. International Journal of Multidisciplinary Research and Development. 2(2):32-36

Watercraft, W.C. (2005). "Order Carnivora". In Wilson, D.E.; Reeder, D.M (eds.). Mammal Species of the World: A Taxonomic and Geographic Reference (3rd ed.). Johns Hopkins University Press. pp. 532-62

21

Insect as Food: Entomophagy

Prince Sahu[1], Aman Pratap Singh[2], Kartikey Singh[1*], Nahida Afreen[1] Paritosh Tripathi[1] and Pankaj Kumar Rajpoot[1]

[1]Department of Entomology, Chandra Shekhar Azad University of Agriculture and Technology, Kanpur, Uttar Pradesh, India

[2]Department of Entomology, Acharya Narendra Dev University of Agriculture and Technology, Ayodhya, Uttar Pradesh, India

Introduction

Entomophagy, the consumption of insects, is rooted in human evolutionary history (Fontaneto, 2011). Throughout Africa, Europe, Asia, and Latin America, insects have held a vital position in the timeline of human nutrition. There exist more than 1900 documented species of insects that serve as human diets at globally, with notable groups encompassing grasshoppers (Orthoptera), caterpillars (Lepidoptera), grubs (Coleoptera), wringed termites (Isoptera), some hymenopteran insects, cicadas (Hemiptera), and a diverse array of aquatic insects. It is intriguing to observe that insects are regularly consumed by over two billion people, and in specific regions, they make up a substantial portion of the animal proteins consumed. (Van Huis, 2013). With its wide adoption and favorable comparison to conventional livestock production in terms of nutritional value and environmental impact, the consumption of insects (entomophagy) has the potential to play a significant role in addressing under nutrition within the increasing global population (Van Huis *et al.*, 2013).

Diversity of edible insect species in the world

Insects have thrived for a remarkable span of time, with a history that dates back at least 400 million years, positioning them as one of the earliest known land-dwelling creatures. More than 390 million years ago, insects diverged as a part of one of the largest subphyla within the arthropod family, undergoing a rapid evolution and radiation that surpasses the pace of any other group (Gaunt and Miles, 2002). In terms of population size, insects hold the title for being the largest animal group on our planet, comprising up to 80% of the animal kingdom. Due to their nutritional significance and abundant availability, over two billion individuals regularly partake in the consumption of edible insects. (Premalatha, 2011).

Various authors have documented the inclusion of over 1900 species of edible insects in the diets of 300 ethnic groups across 113 countries worldwide (Mac Evilly, 2000) (Table 1). In Africa alone, Van Huis *et al.* (2013) reported that 246 species of edible insects have been documented in 27 countries.

Ramos-Elorduy conducted a study two years later, which highlighted Africa as a significant hotspot for the biodiversity of edible insects. The study recorded 524 species from 34 African countries (Ramos-Elorduy, 2005). The majority of these species belong to the orders Orthoptera, Lepidoptera, Coleoptera, Hymenoptera, and Isoptera. Figure 1 provides a checklist of the most commonly consumed edible insect species worldwide, along with their corresponding orders and locations.

Figure 1. Recorded number of edible insect species, by country. Source: Centre for Geo Information, Wageningen University, based on data compiled by Jongema, 2012, Banjo *et al.* 2006, Igwe *et al.*, 2011, Opara *et al.*, 2012, Kelemu *et al.*, 2015.

Why eat insects?

The issue of hunger and malnutrition presents a significant challenge as the global population continues to expand. To ensure food security, it is crucial for the world's food supply to grow at a rate that matches or exceeds population growth. Consequently, there is an ongoing pursuit of new food sources, including the exploration and cultivation of locally specific ethnic food options (Van Huis *et al.*, 2013).

In various regions around the world, particularly in Africa and Latin America, the availability of food resources is becoming increasingly scarce, leading to higher costs of food imports. Therefore, it is crucial to identify and develop local food resources. Researchers have directed their focus towards insects as a potential solution to not only address the challenges posed by rapid population growth but also to other pressing issues. Insects are abundant, possess significant biomass, and offer high-quality protein. Moreover, the longstanding cultural practices in many diverse communities in Africa and Latin America, where insects are consumed in live, roasted, and fried forms, contribute to their appeal as a nutritious protein source that is easily digestible (Conconi, 1974; Wand *et al.*, 2004). Additionally, insects are rich sources of essential nutrients such as fat, vitamins, and minerals, particularly iron and zinc (Oliveira *et al.*, 1976). The multifaceted benefits of insects as food further strengthen their selection as a viable option.

Edible insects as food for humans and feed for animals

It is estimated that around 80% of all species are accounted for by insects, which are recognized as the most abundant and diverse multicellular organisms on our planet (Scaraffia *et al.*, 2012). They play a vital role in pollinating numerous crops, thereby contributing to their growth and reproduction. Moreover, insects provide various agricultural and human health benefits beyond their pollination services (Dzerefos *et al.*, 2013).

There are more than 1900 known species of insects that are incorporated into human diets, with over 2 billion people regularly consuming them (De Foliart, 1989). In certain regions, insect consumption contributes significantly to the intake of animal protein (De Foliart, 1989). Insects continue to hold significant importance as a preferred and vital food source, providing essential protein, fat, minerals, and vitamins, particularly in developing countries and among diverse cultures worldwide (Durst *et al.*, 2010). Comparable nutritional value to meat and fish has been discovered in certain edible insects, while others exhibit even higher proportions of protein, fat, and energy (De Foliart, 1992). The demand for alternative protein sources has gained considerable importance due to fast urbanization in developing nations and shifting global food preferences. Particular significance is held by Coleopterans, Hymenopterans, Isopterans, Lepidopterans, Odonata, and Orthopterans among the various insect orders consumed worldwide, and they are highly regarded. (Fasoranti and Ajiboye, 1993) (Figure 2). Notable examples within these orders include locusts, termites, worms, grasshoppers, caterpillars, palm weevils, and beetle grubs, among others.

Despite being commonly regarded as pests that adversely affect humans, plants, and animal well-being, insects play a crucial role in mitigating food insecurity and providing ecosystem services, including pollination, waste degradation, and biological control. Additionally, insects play a crucial role as a primary food source for a wide variety of animal species. According to Van Huis *et al.* (2013) food and feed security are upheld by importance of insects.

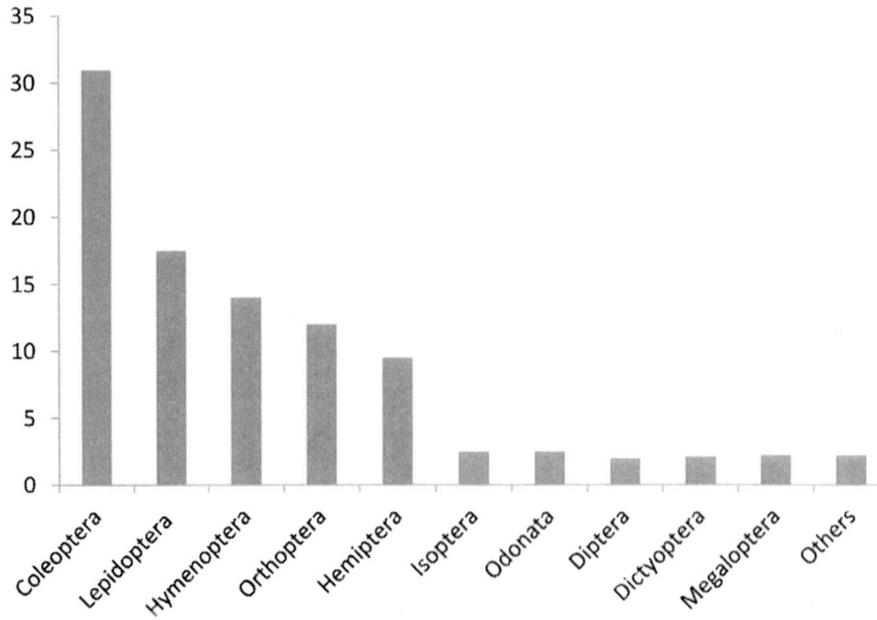

Figure 2. Number of insect species, by order, consumed worldwide. Source: Jongema, 2012.

Importance of edible insects as medicine

The scientific validation of traditional knowledge in bioprospecting has brought greater recognition and importance to edible insects. These insects have long been significant dietary components and remedies for illnesses in various regions worldwide (Srivastava and Gupta, 2009). Traditional healers have successfully utilized insects as medicinal treatments for a range of diseases in both humans and animals. These include conditions such as fever, scabies, epilepsy, severe headaches, bronchitis, hemorrhage, and even dog bites. Insects are also employed in wound healing, gangrene prevention, and stimulating milk production in lactating women, among other applications (Tango, 1981). Modern hospitals are increasingly adopting these traditional insect-based treatments. Furthermore, researchers have exploited the defensive chemicals produced by edible insects to develop antibacterial and anticancer drugs. As an illustration, pierisin, a protein extracted from the pupa of the cabbage butterfly, demonstrates cytotoxic effects on human gastric cancer. Likewise, cecropin has exhibited cytotoxic properties against lymphoma and leukemia cells in mammals (Srivastava and Gupta, 2009). Despite their high nutritional, economic, and medicinal value, edible insects are often overlooked. The acknowledgment and utilization of the benefits these creatures offer to the environment and human health are essential for scientists.

Market value of edible insects

Rural communities of Asia, Africa and South America, there are a deep understanding of the nutritional value and palatability of consuming insects, making them highly sought after. Insects such as crickets, grasshoppers, and locusts are regarded as seasonal delicacies, while giant water beetles find their way into salads, showcasing their culinary appeal (Srivastava and Gupta, 2009). These communities recognize that insects not only provide a valuable source of protein but also supply essential minerals and vitamins, making them a valuable dietary component and enjoyable snacks.

The immense popularity of edible insects has resulted in numerous species becoming highly sought-after commodities in local markets throughout Africa and tropical and semi-tropical regions worldwide (DeFlorita, 2002). In various African regions, such as Zimbabwe, Nigeria, South Africa, Ivory Coast, and Zambia, many families generate a respectable income by selling insects (Balinga *et al.*, 2004). Collected primarily from bushes and farmland by women and children, these insects are processed before being either consumed or sold in local markets, school areas, and town areas. Furthermore, certain insects undergo processing and are exported to local and international shops and restaurants (Chavunduka, 1975). The Capitalization of edible insects presents substantial economic prospects for numerous households in Africa. However, challenges such as poorly understood and poorly organized market chains greatly limit the growth of agribusiness in the region. Documenting the market constraints encountered by farmers in their efforts to diversify production within the edible insect industry is essential.

Edible insects as source of nutrient

The importance of insects in human nutrition should not be underestimated (Sirithon, 2008). There is ample evidence to affirm that insects serve as an exceptionally nutritious and healthful food source, supplying essential nutrients like fats, carbohydrates, vitamins, proteins, amino acids, fibers, and minerals that are vital for both humans and animals. (Vn Huis A, 2013). However, the nutritional composition of edible insects can vary considerably between and within species, influenced by factors such as the insect's metamorphic stage, habitat, diet, as well as the methods of preparation and processing prior to consumption (Rumpold and Schluter, 2013). While several studies have explored the nutritional composition of specific edible insects in countries such as India, the USA, Mexico, and Thailand, there is a scarcity of comprehensive nutritional data on edible insects from around the world (De Foliart, 1992). Insufficient nutritional information on the majority of edible insects can undermine consumer confidence and impede the integration of insect consumption with other food sources. Hence, the objective of this study is to compile comprehensive nutritional data on globally consumed edible insects.

Dietary energy content of edible insects

As the global population continues to grow, over 3.7 billion individuals are currently affected by malnutrition, primarily caused by insufficient protein and energy intake (Olaf and Michael, 2005). Additionally, the availability of new agricultural land is limited, making it challenging to produce an adequate food supply for the growing population. Consequently, a greater number of people are relying on animal protein, which requires substantial resources to produce. Livestock production is particularly costly due to the high energy input compared to the energy output (Mlcek *et al.*, 2014). Moreover, livestock compete with humans for nutrients and energy resources.

The utilization of insects as a protein source has the potential to contribute to insect conservation by promoting habitat protection (Steinfeld *et al.*, 2006). Edible insects are rich in polyunsaturated fat, which serves as a substantial energy source for sustaining life. The energy content of edible insects varies across species and regions, and it has been observed to be notably higher compared to that of livestock and vegetables (Krause and Mahan, 2003). Therefore, edible insects have the potential to efficiently provide the necessary energy for vital functions and the survival of organisms. According to Myers *et al.*, (2008) A vital role is played by insects as they utilize diverse organic sources in nature, recycle organic waste, and provide nutrients to both farm animals and humans.

Protein and amino acid composition of edible insects

Proteins, consisting of amino acids, are essential organic compounds that serve as the building blocks of life. They can be categorized as either essential or nonessential. Protein plays a fundamental role in the activities of all organisms and is involved in the formation of vital substances such as enzymes, hormones, and hemoglobin. Proteins, being the most prevalent biological macromolecules, are found in every cell and across diverse cellular structures. They demonstrate remarkable diversity, encompassing thousands of distinct types that span from small peptides to large polymers with molecular weights reaching the millions. This vast variety can even be found within a single cell. Moreover, proteins exhibit an extensive array of biological functions and serve as the primary outcomes of genetic information pathways. They function as molecular instruments for the expression of genetic information. Hence, ensuring a reliable protein source is of utmost importance to provide energy for both humans and animals.

Insects have the potential to serve as an energy-efficient protein source for humans (Table 2), whether consumed directly or used as food supplements for livestock (De Foliart, 1989). In numerous local communities globally, especially those in the Amazon region, insects contribute approximately 8-70% of their dietary protein intake (Paoletti and Dufour, 2005). Regarding protein content, the order Lepidoptera exhibits the highest level of crude protein, followed by Coleoptera, while the order Hymenoptera contains the lowest protein content. The significant protein content highlights the nutritional significance of edible insects as a valuable dietary component for both humans and animals, offering a potential alternative to expensive animal protein sources that are often lacking in the diets of rural populations in developing countries (Banjo *et al.*, 2006).

Table 2. Crude protein content present in insect order.

Insect order	Stage	Range (% protein)
Coleoptera	Grub & Adults	23–66
Lepidoptera	Caterpillar & Adults	14–68
Hemiptera	Nymphs & Adults	42–74
Hymenoptera	All stage of life	13–77
Odonata	Naiads & Adults	46–65
Orthoptera	Nymphs & Adults	23–65

Source: Xiaoming *et al.*, 2010

Xiaoming *et al.* (2010) conducted an assessment of protein content in 100 species from various insect orders. The study revealed that the protein content was notably high, ranging from 13% to 77% of the dry matter. Additionally, significant variations in protein content were observed both between and within the different insect orders.

Research has indicated that when compared on a mass basis, edible insects contain a higher protein content than other animal and plant-based foods, such as beef, chicken, fish, soybeans, and maize (Teffo *et al.*, 2007). The Food and Agriculture Organization (FAO) conducted a comparison of protein content among insects, reptiles, cattle, and fish (Table 3) (FAO, 2004). It was indicated by the findings that higher protein contents were exhibited by the chosen insect species compared to fish and mammals. This significant finding strongly supports the notion that insects are a valuable protein source, highlighting the importance of promoting their consumption.

Table 3. Comparison of the average protein content among insects, fish, reptiles, and cattle.

Animal group	Species	Edible stage	Protein content (g/100 g fresh weight)
Insect (raw)	Locusts and grasshoppers: Migratory locust (*Locusta migratoria*), Sahelian tree locust *(Anacridium melanorhodon*),edible long-horned grasshopper (*Ruspolia differens*)	Nymph	14-18
		Adult	13-28
	Sphenarium purpurascens (corn-field grasshopper) (chapulines, Mexico)	Adult	35-48
	Silkworm moth (*Bombyx mori*)	Caterpillar	10-17
	Palm worm beetles: The South American palm weevil, African Palm Weevil, flat-faced longhorn beetle	Grub	7-36
	Yellow mealworm	Caterpillar	14-25
	Crickets	Adult	8-25
	Termites	Adult	13-28
Cattles	Turtles: *Chelodina rugosa*, *Chelonia depressa*	Flesh	25-27
		Intestine	18
		Liver	11
		Heart	17–23
		Liver	12–27
Fish (raw)	Finfish	Tilapia	16–19
		Mackerel	16–28
		Catfish	17–28
	Crustaceans	Lobster	17–19
		Prawn (Malaysia)	16–19
		Shrimp	13–27
	Mollusks	Cuttlefish	15–18

Sources: FAO, 2004, 2012

Substantial amounts of amino acids, notably tryptophan and lysine, have been consistently found in edible insects through various studiesSignificant benefits can be offered by supplementing lysine-deficient staple cereals through the incorporation of these insect species into diets. (Elemo *et al.,* 2011). Additionally, many scientific studies have revealed that the insect species Tenebrio molitor contain a significant quantity of essential and nonessential amino acids compared to beef (Table 4).

Table 4. Average amino acid content in *Tenebrio molitor* and beef in g/kg dry matter.

Amino acid	*T. molitor* g/kg dry matter	Beef g/kg dry matter
Essential		
Isoleucine	24.7	16
Leucine	52.2	42
Lysine	26.8	45
Methionine	6.3	16
Phenylalanine	17.3	24
Tryptophan	3.9	-
Semi-essential		
Arginine	25.5	33
Histidine	15.5	20
Methionine+ cysteine	10.5	22
Nonessential		
Alanine	40.4	30
Aspartic acid	40.0	52
Glycine	27.3	24
Glutamic acid	55.4	90
Proline	34.1	28
Serine	25.2	27

Sources: Finke, 2002, Oonincx and Dierenfeld, 2012

Composition of Fat and carbohydrate in edible insects

Fats and carbohydrates are crucial nutrients in the human body, serving as the most important energy source, dropping the need for protein consumption, and aiding in detoxification processes (Alamu *et al.*, 2013). A crucial role is played by edible insects as a valuable carbohydrate source as well. Sting bugs and cicadas are containing carbohydrate 6.71% and 15.98% respectively (Raksakantong *et al.*, 2010). Their carbohydrates are primarily derived from chitin, a macromolecular compound that holds high nutritional and health value. Chitin has been shown to effectively reduce serum cholesterol levels (Burton and Zaccone, 2007). Furthermore, recent reports indicate that edible insects contain substantial amounts of polysaccharides, which have the potential to enhance immune functioning in the human body (Sun *et al.*, 2007). Fat, as the most energy-dense macronutrient in food, is composed of triglycerides and three fatty acid molecules forming its backbone. Fatty acids can be categorized as saturated, unsaturated, or essential. Research from dietary interventions and epidemiological studies has emphasized the significant role of fatty acid intake in human health. Lowering the consumption of dietary saturated fatty acids has been associated with a decrease in factor VII coagulant activity, which is considered a risk factor for cardiovascular disease (Lefevre *et al.*, 2004). Scientific literature further indicates that the consumption of polyunsaturated fatty acids (PUFAs) offers various health benefits for humans. These benefits include the reduction of glucose tolerance, thereby decreasing the risk of diabetes and blood pressure (Sirtori and Galli, 2002), prevention of insulin resistance (Mori *et al.*, 1997), inhibition of thromboxane A2 formation, thus reducing the tendency for blood clotting (Nieuwenhuys and Hornstra, 1998), & lowering blood pressure (Asaf *et al.*, 2002). Womeni *et al.* (2009) revealed that content and composition of fat extracted from various insect species (Table 5).

Their findings revealed that the oils extracted from insects are abundant in polyunsaturated fatty acids, often including the essential fatty acids linoleic acid and α-linolenic acid. These two essential fatty acids are widely acknowledged for their nutritional significance, particularly in promoting the healthy development of children and infants (Michaelsen *et al.*, 2009).

Table 5. Fat content and randomly selected fatty acids of several edible insect species consumed in Cameroon

Sr. No.	Edible insect species	Composition of main fatty acids (% of oil content)	Type of fatty acid
1	African palm weevil (*Rhynchophorus phoenicis*)	Palmitoleic acid 38%	Monounsaturated fatty acids
		Linoleic acid 45%	polyunsaturated fatty acids
2	Edible grasshopper (*Ruspolia differens*)	Palmitoleic acid 28%	Monounsaturated fatty acids
		Linoleic acid 46%	polyunsaturated fatty acids
		a-Linolenic acid	polyunsaturated fatty acids
3	Termites (*Macrotermes* sp.)	Palmitoleic acid	saturated fatty acids
		Oleic acid	Monounsaturated fatty acids
		Stearic acid 9%	saturated fatty acids
4	Saturniid caterpillar (*Imbrasia* sp.)	Palmitic acid 8%	Monounsaturated fatty acids
		Oleic acid 9%	polyunsaturated fatty acids
		Linoleic acid 7%	polyunsaturated fatty acids
		a-Linolenic acid 38%	Monounsaturated fatty acids
5	Variegates grasshopper (*Zonocerus variegates*)	Palmitoleic acid 24%	polyunsaturated fatty acids
		Linoleic acid 45%	polyunsaturated fatty acids
		a-Linolenic acid	polyunsaturated fatty acids
		g-Linolenic acid	polyunsaturated fatty acids
		Oleic acid 11%	Monounsaturated fatty acids

Source : Womeni *et al.*, 2009

Mineral composition of edible insects

Minerals have significant roles in the metabolic and physiological functions of living organisms. Certain minerals such as iron, zinc, copper, and manganese serve as antioxidants and enzyme cofactors, thereby strengthening the immune system (Talwar *et al.*, 1989). Similarly, minerals like magnesium, zinc, and selenium help prevent conditions such as cardiomyopathy, muscle degeneration, growth retardation, impaired spermatogenesis, immunological dysfunction, and bleeding disorders (Chaturvedi *et al.*, 2004). Iron deficiency poses a major concern in the diets of women, especially pregnant women, particularly in developing regions, particularly in Africa (Orr, 1986). Magnesium, on the other hand, is involved in over 300 biochemical reactions within the body. It aids in maintaining normal muscle and nerve function, regulates the heart rhythm, and helps regulate blood sugar levels (Saris *et al.*, 2000).

Table 6. Mineral and ash (mg/100 g) contents in some edible insect species.

Insect species	Ca	P	Fe	Mg	Ash
leafcutter bees	1.00	14.90	9.56	60.96	7.60
Macrotermes bellicosus	21	136	27	0.15	2.90
banded garden spider	61.28	136.4	18.2	6.14	4.21
Anaphe infracta	8.56	111.3	1.78	1.01	1.60
Annona reticulata	10.52	102.4	2.24	2.56	2.50
Lepidoptera litoralia	12.00	9.00	19.50	0.50	4.30
A. venata	8.57	100.5	2.01	1.56	3.20
A. mellifera	15.4	125.5	25.2	5.23	2.20
Oryctes monoceros	NA	NA	85.00	175.00	10.50
Gymnelia lucens	NA	NA	NA	NA	6.40
Aphodius rufipes	42.16	131.2	30.82	11.72	2.74

NA, not available. Sources: Banjo *et al.* 2006

The analysis of elemental composition in edible insects has revealed their richness in essential nutrients such as potassium and sodium (cricket nymph), calcium (adult cricket), copper (mealworm), iron (axayacatl), zinc (cricket), manganese (cricket), and phosphorus. This indicates that edible insects have the potential to provide the necessary nutritional elements required for optimal human body functions. By including edible insects in the di*et al*ong with other nutrient-rich foods and animals, the overall nutritional profile can be enhanced. This approach ensures a diverse intake of essential minerals. Table 6 presents the mineral content of specific edible insects found in Nigeria.

Vitamin composition present in edible insects

Vitamins, essential organic compounds, have a crucial role in human metabolism. Unlike other substances, the human body cannot synthesize vitamins internally, making it necessary to consistently obtain them from dietary sources (Alamu, 2013). Ascorbic acid, also known as vitamin C, functions as an antioxidant and reducing agent, while components of coenzymes make up the vitamin B group. Vitamin K is necessary for normal blood clotting, and vitamin A is crucial for maintaining proper vision. Numerous studies have demonstrated that edible insects contain significant amounts of vitamins (Ekpo, 2005). Edible insects, with their significant vitamin content, show great promise as food supplements for malnourished individuals and animals. Table 7 presents the vitamin contents of various insect orders, all of which contain noteworthy levels of vitamins A, B_2, and C.

Table 7. Vitamin (mg/100 g) contents in some edible insects in Nigeria.

Order	N	Vitamin A		Vitamin B_2		Vitamin C	
		Range	Mean	Range	Mean	Range	Mean
Isoptera	3	0.026–0.05	0.14	1.54–1.98	1.69	3.01–17.76	8.06
Coleoptera	3	0.086–0.125	0.11	0.08–2.62	1.64	4.25–7.59	5.75
Orthoptera	3	0.0–0.068	0.03	0.03–0.08	0.06	0.0–8.64	3.21
Lepidoptera	5	0.028–0.034	0.09	0.09–2.21	1.50	1.95–4.52	2.83
Hymenoptera	1				3.24		10.25

N, number of insect species in each order; Sources: Banjo *et al.*,2006; Ifie and Emeruwa, 2011; Igwe *et al.*, 2008

Opportunities for insect rearing in the environment (ecological opportunities)

Insects provide a multitude of ecological services that are vital for the survival of humanity. The utilization of insects as a protein source can contribute to insect conservation through habitat protection (Steinfeld *et al.*, 2006). Insects play a crucial role as they feed on organic matter in nature, efficiently utilizing all available organic sources. They also play a significant role in recycling organic waste and providing nutrients for farm animals (Myers *et al.*, 2008, Diener *et al.*, 2009). Many insect species are essential for improving soil fertility. They aid in the breakdown of waste products, making them suitable for consumption by fungi and bacteria. As a result of this decomposition process, minerals and nutrients are released, making them readily available in the soil for plants to absorb. This, in turn, improves soil fertility. For example, animal carcasses are consumed by fly maggots and beetle larvae, and the decomposition of manure is significantly contributed to by dung beetles, with around 400 known species. Consequently, insects can serve as efficient bio-transformers, converting abundant and cost-effective organic waste into animal biomass rich in proteins, suitable for animal nutrition (FAO, 2009).

Beneficial roles of insects in human welfare

In addition to being sources of food, edible insects offer a wide range of valuable products to humans. Numerous insect species possess significant commercial and pharmaceutical value. For instance, bees are known for producing substantial quantities of honey, while silkworms yield vast amounts of silk, both of which can be sold in local and international markets (FAO, 2009). Silkworms alone produce over 90,000 tons of silk (Youn-woo, 1999). Scale insects of the Hemiptera order produce carmine, a red dye used in coloring foods, textiles, and pharmaceuticals. Potential has been shown in medical applications for repairing arteries using resilin, a rubber-like protein that enables insects to jump, owing to its elastic properties. (Elvin *et al.*, 2005). Moreover, other products derived from edible insects, like honey bees wich produced honey, propolis, royal jelly, and venom, have been used in the treatment of traumatic and infected wounds and burns (Van Huis A., 2003). Additionally, insect products have found applications in engineering methods for the production of biomaterials (Lewis, 1992).

Conclusion

Meeting global food demands in a sustainable manner presents one of the greatest challenges for humanity, and it has garnered significant attention in recent years (West *et al.*, 2014). It is widely acknowledged that food security is significantly bolstered by agriculture, which enhances the availability of economical food and uplifts the lives of the underprivileged. Within the framework of a sustainable diet, the integration of insects as a source of food and feed assumes a crucial role in ensuring food security and enhancing the well-being of individuals in Africa. They are rich in protein and amino acids, including essential amino acids that are essential for maintaining human health. Additionally, they offer a valuable source of unsaturated fatty acids, minerals, vitamins, and carbohydrates, providing exceptional nutritional value. Moreover, their significance extends to medical, commercial, and ecological aspects. Given these remarkable attributes, edible insects warrant thorough consideration as a means to address the pressing global challenge of human nutrition.

References

Alamu, O.T., Amao, A.O., Nwokedi, C.I., Oke, O.A. and Lawa, I.O. (2013). Diversity and nutritional status of edible insects in Nigeria: a review. International Journal of Biodiversity and Conservation, 5(4): 215-222

Balinga, M.P., Mapunzu, P.M., Moussa, J.B. and N'gasse, G. (2004). Contribution of forest insects to food security. The example of caterpillars in Central Africa

Banjo, A.D., Lawal, O.A. and Songonuga, E.A. (2006). The nutritional value of fourteen species of edible insects in southwestern Nigeria. African journal of Biotechnology, 5(3): 298-301

Bodenheimer, F.S. (1951). Insects as human food. The Hague: Dr. W.

Burton, O.T. and Zaccone, P. (2007). The potential role of chitin in allergic reactions. Trends in Immunology, 28(10): 419-422

Chaturvedi, U.C., Shrivastava, R. and Upreti, R.K. (2004). Viral infections and trace elements: a complex interaction. Current Science, 1536-1554

Chavunduka, D.M. (1975). Insects as a source of protein to the African. Rhodesia Sci. News, 9: 217-220

Conconi, J.R.E. (1974). Insects as a source of proteins in the future (prospective). Reg. Sec. Educ. Pub, 1639: 74

DeFoliart, G.R. (2002). The human use of insects as a food resource: a bibliographic account in progress. University of Wisconsin, Madison, WI

DeFoliart, G.R. (1989). The human use of insects as food and as animal feed. Bull. Entomol. Soc. Am., 35: 22–35

DeFoliart, G.R. (1992). Insects as human food nutritional and economic aspects. Crop Protect., 11: 395–399

Diener, S., Zurbrügg, C. and Tockner, K. (2009). Conversion of organic material by black soldier fly larvae: establishing optimal feeding rates. Waste Management & Research, 27(6): 603-610

Durst, P.B. and Shono, K. (2010). Edible forest insects: exploring new horizons and traditional practices. Forest insects as food: humans bite back, 1-4

Dzerefos, C.M., Witkowski, E.T.F. and Toms, R. (2013). Comparative ethno-entomology of edible stinkbugs in southern Africa and sustainable management considerations. Journal of Ethnobiology and Ethnomedicine, 9(1): 1-12

Ekpo, K.E. and Onigbinde, A.O. (2005). Nutritional potentials of the larva of Rhynchophorus phoenicis (F). Pakistan Journal of Nutrition, 4(5): 287-290

Elemo, B.O., Elemo, G.N., Makinde, M.A. and Erukainure, O.L. (2011). Chemical evaluation of African palm weevil, Rhychophorus phoenicis, larvae as a food source. Journal of Insect Science, 11(1): 146

Elvin, C.M., Carr, A.G., Huson, M.G., Maxwell, J.M., Pearson, R.D., Vuocolo, T. and Dixon, N.E. (2005). Synthesis and properties of crosslinked recombinant pro-resilin. Nature, 437(7061): 999-1002

FAO (2004). Contribution of forest insects to food security. The example of the Central African caterpillars (available at: www.fao.org/docrep/007/j3463e/j346 3f00.html)

FAO (2009). The State of Food and Agriculture: Livestock in the Balance.. Rome: Food and Agriculture Organization of the United Nations (FAO)

Fasoranti, J.O. and Ajiboye, D.O. (1993). Some edible insects of Kwara state, Nigeria. American Entomologist, 39(2): 113-116

Finke, M.D. (2002). Complete nutrient composition of commercially raised invertebrates used as food for insectivores. Zoo biology: published in affiliation with the American zoo and aquarium association, 21(3): 269-285

Fontaneto D., Tommaseo-Ponzetta M., Galli C., Risé P., Glew R.H. and Paoletti M.G. (2011). Differences in fatty acid composition between aquatic and terrestrial insects used as food in human nutrition. Ecology of Food and Mutrition, 50(4): 351-367

Gaunt, M.W. and Miles, M.A. (2002). An insect molecular clock dates the origin of the insects and accords with palaeontological and biogeographic landmarks. Molecular Biology and Evolution, 19(5): 748-761

Ifie I. and Emeruwa C.H. (2011). Nutritional and anti-nutritional characteristics of the larva of Oryctes monoceros. Agric. Biol. JN Am, 2(1): 42-46

Igwe C.U., Ujowundu C.O., Nwaogu L.A. and Okwu, G.N. (2011). Chemical analysis of an edible African termite Macrotermes nigeriensis, a potential antidote to food security problem. Biochemistry and Analytical Biochemistry, 1(105): 2161-1009

Jongema, Y. (2012). List of edible insect species of the world. Wageningen, Laboratory of Entomology, Wageningen University

Kelemu, S., Niassy, S., Torto, B., Fiaboe, K., Affognon, H., Tonnang, H. and Ekesi, S. (2015). African edible insects for food and feed: inventory, diversity, commonalities and contribution to food security. Journal of Insects as Food and Feed, 1(2): 103-119

Krause M., Mahan L.M. Food, Nutrition and Diet Therapy (2003). 11th ed. St. Louis, CA, USA: W. B. Saunders Co.

Lee, Y.W. (1999). Silk reeling and testing manual (No. 136). Food & Agriculture Organization

Lefevre, M., Kris-Etherton, P.M., Zhao, G. and Tracy, R.P. (2004). Dietary fatty acids, hemostasis, and cardiovascular disease risk. Journal of the American Dietetic Association, 104(3): 410-419

Lewis, S.E. (1992). Insects of the Klondike Mountain Formation, Republic, Washington. Washington Geology, 20(3): 15-19

MacEvilly, C. (2000). Bugs in the system. Nutr. Bull., 25: 267-268

Michaelsen K.F., Lauritzen L. and Mortensen E.L. (2009). Effects of breast-feeding on cognitive function. Breast-feeding: early influences on later health, 199-215

Mlček, J., Rop, O., Borkovcova, M. and Bednářová, M. (2014). A comprehensive look at the possibilities of edible insects as food in Europe-a review. Polish Journal of Food and Nutrition Sciences

Mori, Y., Murakawa, Y., Katoh, S., Hata, S., Yokoyama, J., Tajima, N. and Shibutani, Y. (1997). Influence of highly purified eicosapentaenoic acid ethyl ester on insulin resistance in the Otsuka Long-Evans Tokushima Fatty rat, a model of spontaneous non-insulin-dependent diabetes mellitus. Metabolism, 46(12): 1458-1464

Moudgil, K.D. and Talwar, G.P. (1989). Text book of biochemistry and human biology. Prentice Hall of India

Müller, O. and Krawinkel, M. (2005). Malnutrition and health in developing countries.Cmaj, 173(3): 279-286

Myers H.M., Tomberlin J.K., Lambert B.D. and Kattes D. (2014). Develop-ment of black soldier fly (Diptera: Stratiomyidae) larvae fed dairy manure. Environmental Entomology, 37(1): 11-15

Nieuwenhuys, C.M. and Hornstra G. (1998). The effects of purified eicosapentaenoic and docosahexaenoic acids on arterial thrombosis tendency and platelet function in rats. Biochimica et Biophysica Acta (BBA)-Lipids and Lipid Metabolism, 1390(3): 313-322

Oliveira, J.S., de Carvalho, J.P., De Sousa, R.B. and Simao, M.M. (1976). The nutritional value of four species of insects consumed in Angola. Ecology of Food and Nutrition, 5(2): 91-97

Oonincx D.G.A.B. and Dierenfeld E.S. (2012). An investigation into the chemical composition of alternative invertebrate prey. Zoo Biology, 31(1): 40-54

Opara M.N., Sanyigha F.T., Ogbuewu I.P. and Okoli I.C. (2012). Studies on the production trend and quality characteristics of palm grubs in the tropical rainforest zone of Nigeria. Journal of Agricultural Technology, 8(3): 851-860

Orr B. (1986). Improvement of women's health linked to reducing widespread anemia. Int. Health News, 7(3): 56-68

Paoletti M.G., Buscardo E. and Dufour D.L. (2000). Edible invertebrates among Amazonian Indians: a critical review of disappearing knowledge. Environment, Development and Sustainability, 2: 195-225

Premalatha, M., Abbasi, T., Abbasi, T. and Abbasi, S.A. (2011). Energy-efficient food production to reduce global warming and ecodegradation: The use of edible insects. Renewable and sustainable energy reviews, 15(9): 4357-4360

Qureshi, A.A., Sami, S.A. and Khan, F.A. (2002). Effects of stabilized rice bran, its soluble and fiber fractions on blood glucose levels and serum lipid parameters in humans with diabetes mellitus Types I and II. The Journal of nutritional biochemistry, 13(3): 175-187

Raksakantong, P., Meeso, N., Kubola, J. and Siriamornpun, S. (2010). Fatty acids and proximate composition of eight Thai edible terricolous insects. Food Research International, 43(1): 350-355

Ramos-Elorduy J. (2005). Insects a hopeful food. In: Paoletti M, editor. Ecological Implications of Mini livestock for Sustainable Development. USA: Science Publishers Inc:. pp. 263–29

Rumpold, B. A. Schlüter, oK (2013). Nutritional composition and safety aspects of edible insects. Molecular Nutrition and Food Research, 57(3)

Saris, N.E.L., Mervaala, E., Karppanen, H., Khawaja, J.A. and Lewenstam, A. (2000). Magnesium: an update on physiological, clinical and analytical aspects. Clinica chimica acta, 294(1-2): 1-26

Scaraffia, P.Y. and Miesfeld, R.L. (2012). Insect Biochemistry/Hormones. BCH2 00093. University of Arizona, Tuczon, AZ, USA

Siriamornpun, S. and Thammapat, P. (2008). Insects as a delicacy and a nutritious food in Thailand. Using Food Science and Technology to Improve Nutrition and Promote National Development, 16, 16

Sirtori C.R., Galli C. (2002). N-3 fatty acid and diabetes. Biomed. Phamacother, 56: 397-406

Srivastava, J.K. and Gupta, S. (2009). Health promoting benefits of chamomile in the elderly population. In Complementary and Alternative Therapies and the Aging Population (pp. 135-158). Academic Press

Steinfeld, H. (2006). Livestock's long shadow: environmental issues and options. Food & Agriculture Org

Sun L., Feng Y., He Z., Ma T., Zhang X. (2007). Studies on alkaline solution extraction of polysaccharide from silkworm pupa and its immunomodulating activities. Forest Res.; 20: 782-786

Tango M. (1994). Les Insectes CommeAliments de LHomme. Published in 1981, reviewed in Food Insects Newslett 7(3): 3–4

Teffo, L.S., Toms, R.B. and Eloff, J.N. (2007). Preliminary data on the nutritional composition of the edible stink-bug, Encosternum delegorguei Spinola, consumed in Limpopo province, South Africa. South African Journal of Science, 103(11-12): 434-436

Van Huis, A. (2003). Insects as food in sub-Saharan Africa. International Journal of Tropical Insect Science, 23(3): 163-185

Van Huis, A., Van Itterbeeck, J., Klunder, H., Mertens, E., Halloran, A., Muir, G. and Vantomme, P. (2013). Edible insects: future prospects for food and feed security (No. 171). Food and agriculture organization of the United Nations

Wang, D., Bai, Y. Y., Li, J. H. and Zhang, C. X. (2004). Nutritional value of the field cricket (Gryllus testaceus Walker). Insect Science, 11(4): 275-283

West, P.C., Gerber, J.S., Engstrom, P.M., Mueller, N.D., Brauman, K.A., Carlson, K.M. and Siebert, S. (2014). Leverage points for improving global food security and the environment. Science, 345(6194): 325-328

Womeni, H.M., Linder, M., Tiencheu, B., Mbiapo, F.T., Villeneuve, P., Fanni, J. and Parmentier, M. (2009). Oils of insects and larvae consumed in Africa: potential sources of polyunsaturated fatty acids. Oléagineux, Corps gras, Lipides, 16(4-5-6): 230-235

Xiaoming, C., Ying, F., Hong, Z. and Zhiyong, C. (2010). Review of the nutritive value of edible insects. Forest insects as food: humans bite back, 85

Ying, F. E. N. G., Zhao, H. E., Tao, M. A. and Xin, Z. H. A. N. G. (2007). Studies on Al ka line Solution Extraction of Polysacchar idefrom Sil kworm Pupa and Its Immunomodula ting Activ ities. 20(6): 782-786

22

Diseases of Honey Bees

Kartikey Singh[1], Nahida Afreen[1*], Prince Sahu[1], Aman Pratap Singh[2] and Saurabh Singh[3]

[1]*Department of Entomology, Chandra Shekhar Azad University of Ag. and Tech. Kanpur-208002, Uttar Pradesh, India*

[2]*Department of Entomology, Acharya Narendra Deva University of Agriculture and Technology, Kumarganj, Ayodhya-224229, Uttar Pradesh, India*

[3]*Department of Entomology, Banda University of Agriculture and Technology Banda-210001, Uttar Pradesh, India*

Introduction

Bees belonging to the Apis genus can be found in a wide range of climates worldwide. Among them, the *Apis mellifera* Linnaeus species is distributed across diverse environments such as sub-Saharan Africa, northern Europe, and Central Asia. This species exhibits remarkable adaptability, thriving in various habitats like the oases of the African desert, the Alps, the edges of the tundra, and the misty regions of the United Kingdom. Additionally, other honey bee species within the Apis genus are predominantly distributed in Asia, specifically in tropical Southeast Asia.Honeybees are good biological indicators. Honey bees (*Apis mellifera* L.) are insects of great importance for humanity. Humans have taken advantage from them from ancient times, both directly –through beekeeping– and indirectly –through farming, so their importance goes beyond ecology, being also economic.

Honey bees and honey bee health have become a major topic recently due to the important role honey bees play in pollination and food production. Although it is often suggested that mankind will not survive for long once honeybees are gone, this is rather exaggerating the role of honey bees for human nutrition since primary food production, and especially our staple foods, is independent of animal (insects, birds, bats) pollination. Our staple foods (e.g. wheat, corn, rice, potatoes) are wind- or passively self-pollinated or are vegetatively propagated, meaning that their production does not increase with animal pollinators.90% commercial pollination is performed by managed honey bees (*Apis mellifera* L.).

Honey bees are susceptible to a variety of diseases which are follows:

European foulbrood (EFB)

EFB is a bacterial brood disease that affects honey bee colonies, caused by the bacterium *Melissococcus plutonius*. They affect the brood (larvae and pupae) of honey bee colonies.Here are the typical symptoms of European foulbrood disease in honey bees:

- Infected brood cells appear sunken and may have a color ranging from yellow to brown. The larvae inside the cells may be discolored and have a foul-smelling odor.
- The capping of infected brood cells may become perforated due to the decay of the underlying larvae.
- Infected brood frames may show a spotty brood pattern with healthy and infected cells intermixed.

Management

- Regularly inspect the hives for signs of infection. Look for symptoms such as sunken and perforated cappings, discolored larvae, and foul odor coming from the brood cells.
- If you find an infected hive, isolate it from other colonies to prevent the spread of the disease. Move it to a separate location or at least several hundred meters away from healthy colonies.
- Encourage hygienic behavior in bees.
- Oxytetracycline and tylosin are commonly used as antibiotics for control European foul brood.
- Sterilize hive tools and equipment between inspections to prevent the transmission of bacteria between colonies.

American foulbrood (AFB)

AFB is a highly destructive bacterial infection that primarily targets the brood, which includes both larvae and pupae, within honey bee colonies. American foulbrood is the result of an infection caused by *Bacillus larvae*, which is a bacterium characterized by its rod-shaped morphology measuring approximately 2.5-5 μm in length and 0.5-0.8 μm in width. This bacterium is capable of movement through peritrichous flagella and exhibits a Gram-positive staining pattern. Let's explore the characteristic signs of American foulbrood disease observed in honey bee larvae:

- American foulbrood is a condition that affects larvae, typically leading to their demise after they have completed their cocoon formation and lie sprawled on their backs, oriented with their heads towards the cap of the cell. Although most commonly observed in propupae, some pupae are also subject to fatality.

- These affected larvae undergo a transformation into a brown hue, undergoing putrefaction and emitting an unpleasant odor reminiscent of fish glue. Over the course of about a month, they desiccate into a firm, adherent scale.
- The typical period for an infected larva to display symptoms of the disease is approximately 12.5 days post-hatching, with nearly all afflicted larvae manifesting visible discoloration within a range of 10 to 15 days post-hatching.
- The caps that cover these diseased larvae rapidly become damp and assume a dark coloration. This causes the caps to depress inward, prompting adult bees to initiate their removal. The process begins with the formation of small openings, eventually culminating in the complete exposure of the cell.
- At the stage where the caps have sunken, if a matchstick is inserted into the remains of the larva and then withdrawn, it extracts the brown, partially liquid remnants in a stringy thread-like manner.

Management

- Thoroughly sanitize previously used equipment.
- Introduce queens from colonies with tolerance to re-queen the hives.
- Provide colonies with supplementary food during periods without nectar availability.
- Position colonies in an appropriate habitat.
- Conduct an early and precise diagnosis of diseases.
- Utilize the shaking technique as an efficient approach to manage American foulbrood (AFB). This method entails relocating adult bees to a hive devoid of drawn combs and eliminating the brood combs within the infected colony. Subsequently, the transferred bees consume the contaminated honey present in their honey stomachs, all the while constructing new combs to replace the ones discarded.

Chalk brood

Chalkbrood is an ailment that impacts the offspring of *Apis mellifera* L. bees. This fungal infection is widespread globally, with a predominant presence in the temperate areas of the northern hemisphere. It is induced by the fungus *Ascosphaeraapis*. This condition impacts all types of honeybee brood, infecting them when they consume food tainted with spores of the *Ascosphaeraapis* fungus.

Symptoms

- Colony health issues become evident through the presence of lifeless brood remains, referred to as mummies, located in front of the hive entrance or on the ground below it.
- Inspecting the combs reveals perforated cappings, with both mummified broods and empty cells, which points to the diligent removal of deceased brood through the hygienic behavior of worker bees.
- Infected larvae typically succumb within the initial 48 hours after capping, during their pupal phase.
- Shortly following their demise, they undergo expansion and become enveloped by the fungus's white mycelia.
- In the subsequent phase of the disease, certain pupae maintain a pale hue reminiscent of chalk (hence the term "chalkbrood"). Others undergo a transformation from brown to gray and, ultimately, black. This discoloration occurs due to the emergence of fruiting bodies, which can vary in size and color.

Management

- As an initial approach to managing the issue, it is advisable to replace aged honeycombs or those displaying signs of A. apis infection in order to eradicate spores from the hive. The introduction of fresh wax into the combs offers the benefit of diminishing spore presence while also stimulating the queen to lay eggs, thereby fostering an increase in the bee population.
- A suggested practice involves an annual replacement of one third of the hive's combs, alongside the meticulous cleansing and sanitization of tools like chisels, gloves, and fumigators subsequent to handling affected hives. Hives exhibiting indications of any disease should be managed in a final phase, reducing the potential for healthy colonies to encounter contamination through the beekeeper's equipment.
- Numerous natural compounds have been evaluated for their antifungal attributes. Essential oils containing citral, geraniol, and citronellal have exhibited notable inhibitory effects on fungal growth in laboratory conditions. In Canada, field studies have been conducted on a wide-ranging antimicrobial substance, lysozyme, showing promising outcomes in the context of chalkbrood control within bee colonies.

Stone Brood

Stonebrood is an exceptionally uncommon malady affecting honey bee broods, provoked by various fungi belonging to the Aspergillus genus. Massen (1906) initially documented this ailment, and it has subsequently been detected across the globe. The most commonly noted species is Aspergillus flavus, succeeded by Asp. fumigatus; however, other species like Asp. niger are also capable of inducing honey bee mortality. Aspergillus can infiltrate the host through ingestion of spores, traversing the gut, or via the outer covering, known as the cuticle. Consequently, adults, as well as larvae and pupae, are susceptible to infection.Furthermore, the majority of Aspergillus varieties produce aflatoxins, which have been proposed as the principal cause of demise among honey bees affected by stonebrood.The indications commonly identified within a colony afflicted by stonebrood bear resemblances to chalkbrood symptoms and encompass irregular sealing of the brood cells. Infected brood, often referred to as "mummies," become discernible within the combs. These mummies undergo a hardening process, adopting a resemblance to petite stones, as opposed to the spongy texture characterizing chalkbrood mummies.

Nosema disease

Nosema ceranae is a type of microsporidian parasite that has been identified as infecting both the Asian honey bee, *Apis cerana*, and the European honey bee, *Apis mellifera*. Among the microsporidian parasites documented within honey bees, only two have been recognized so far, namely *Nosema apis* Zander, 1909 and *Nosema ceranae* Fries *et al.*, 1996. *Nosema apis* was originally identified within the European honey bee (*Apis mellifera* L) and stands as one of the initial microsporidia to be documented (Zander, 1909). To address this ailment, strategies such as securing sufficient winter supplies, practicing hygienic feeding and watering methods, and routinely disinfecting hives using substances like Formidol and the antibiotic fumagillin have been proposed for disease management.

Conclusion

The significance of honey bees to humanity is emphasized, both economically and ecologically, especially in their role as pollinators. Honey bees (*Apis mellifera*) play a crucial role in commercial pollination, contributing to around 90% of the process. The health of honey bees and the control of diseases like European foulbrood, American foulbrood, chalk brood, stone brood, and Nosema disease involve distinct symptom identification and specific management approaches for each ailment. Safeguarding honey bee populations and managing their diseases is not just a matter of ecological conservation; it is essential for our food security, economic stability, and overall environmental health. The survival and well-being of honey bees are intricately linked to our own well-being, making their protection a global priority.

References

Bailey, L. and Ball, B. V. (1991). Honey Bee Pathology (2nd ed.). Academic Press. (ISBN: 978-0120730316)

Berry, J.A., Delaplane, K.S. (2001). Effects of comb age on honeybee colony growth and brood survivorship. Journal of Apicultural Research, v.40, n.1, p. 3-8

Brodsgaard C. J., Hansen H. (1999). Prevention and control of American foulbrood without use of antibiotics. Proc. 36th Int. Api. Congr. Vancouver, 47-48

Burnside, C.E. (1930) Fungous diseases of the honey bee. US Department of Agriculture Technical Bulletin 149

Calderone, N.W., Shimanuki, H., Allen-Wardell, G. (1994). An in vitro evaluation of botanical compounds for the control of the honeybee pathogens Bacillus larvae and Ascosphaeraapis, and the secondary invader B. alvei. J. Essent. Oil Res. 6: 279–287

Castagnino, G.L.B., Funari, S.R., Blume, E., Arboitte, M.Z., Weber, M.N. DoençaCriaGiz Ascosphaera apis (Maassenex Claussen) Olive & Spiltoiremabelhas Apis mellifera L. naDepressão Central do Rio Grande do Sul. Ciência Rural, v.36, n.6, p.1909-1911

Celli G. (1994). L'ape come indicatore biologico dei pesticidi. In, Atti del convegno: "L'ape come insetto test dell'inquinamento agricolo" P.F. "Lotta biologica e integrata per la diffuse delle colture agrarie e delle piante forestali" March 28, 1992, Florence, Italy. (D'Ambrosio, M.T and Accorti, M. Eds.) Ministero Agricoltura e Foreste, Rome, Italy, pp. 15-20

De Graaf, D. C., Alippi, A. M., Antúnez, K., Aronstein, K. A., Budge, G., De Koker, D., de Smet, L., Dingman, D. W., Evans, J. D., Foster, L. J., Fünfhaus, A., Garcia-Gonzalez, E., Gregorc, A., Human, H., Murray, K. D., Nguyen, B. K., Poppinga, L., Spivak, M., vanEngelsdorp, D., Genersch, E. (2013). Standard methods for American foulbrood research. Journal of Apicultural Research, 52(1): 1–28. doi:10.3896/IBRA.1.52.1.14

DE Jong, D. (1976) Experimental enhancement of chalk brood infections. Bee World, v.57, n.3, p.114-115

Evans, J.D., Aronstein, K., Chen, Y.P., Hetru, C., Imler, J.L., Jiang, H., Kanost, M., Thompson, G.J., Zou, Z., Hultmark, D. (2006), Immune pathways and defence mechanisms in honeybees Apis mellifera. Insect Mol. Biol. 2006, 15: 645-656

Flores, J.M., Funari, S.R.C., Ruiz, J.A., Ruz, J.M., Puerta, F., Campano, F. Ascosferiose (Ascosphaeraapis): causaspredisponentes, medidas de controle e prevenção. Boletim de Indústria Animal, v.57, n.2, p.201-209, 2000

Fries, I., Feng, F., D.A. Silva, A., Slemenda, S.B., Pieniazek, N.J. (1996). Nosema ceranae n. sp. (Microspora, Nosematidae), morphological and molecular characterization of a microsporidian parasite of the Asian honey bee Apiscerana (Hymenoptera, Apidae). European Journal of Protistology 32: 356–365

Genersch E., von der Ohe W., Kaatz H., Schroeder A., Otten C., Büchler R., Berg S., Ritter W., Mühlen W., Gisder S., Meixner M., Liebig G., Rosenkranz P. (2010). The German bee monitoring project: a long term study to understand periodically high winter losses of honey bee colonies. Apidologie. doi:10.1051/apido/2010014

Ghazoul J. (2005a). Buzziness as usual? Questioning the globalpollination crisis. Trends EcolEvol 20: 367–373

Gilliam, M., Lorenz, B.J., Richardson, G.V. (1988a). Digestive enzymes and microorganisms in honey bees, Apis mellifera – influence of streptomycin, age, season and pollen. Microbios 55: 95-114

Gilliam, M., Vandenberg, J. (1997). Fungi. In Morse, R.A., Flottum, K. (Eds). Honey bee pests, predators, & diseases (3rd Edition). A I Root Co.; Medina, USA. pp. 81-110

Haseman, L. (1961). How long can spores of American foulbrood live? American Bee Journal, 101: 298-299

Heath, L.A.F. (1982). Development of chalk brood in a honey bee colony: a review. Bee World, v.63, n.3: p.119-130

Katznelson, H., Jamieson, C.A. (1952). Control of nosema disease with fumagillin. Science 115: 70-71

Klein A.M., Vaissiere B.E., Cane J.H., Steffan-Dewenter I., Cunningham S.A., Kremen C., Tscharntke T. (2007). Importance of pollinators in changing landscapes for world crops. Proc R Soc B 274: 303–313

Moreno, C. (2016). Importance and health of honey bee (Apis mellifera): a review. Scientific Writing and Presentations in English, 1-15

Morse, R.A., Flottum, K. (1997). Honey Bee Pests, Predators, and Diseases. A.I. Root Company. (ISBN: 978-0915698013)

Muñoz, I., Cepero, A., Pinto, M.A., Martín-Hernández, R., Higes, M., De la Rúa, P. (2014) Presence of Nosema ceranae associated with honeybee queen introductions. Infect. Genet. Evol. 23: 161-168

Panasiuk, B.; Bieńkowska, M.; Gerula, D.; Węgrzynowicz, P. (2014) Susceptibility of bee larvae to chalkbrood in relation to hygienic behaviour of worker bees in colonies of chosen races of honeybee (Apis mellifera). Journal of Apicultural Science, v. 58, n. 1: p. 119-126

Park, O.W. (1953). Determination of the incubation period of Bacillus larvae White, causal organism of American foulbrood in honeybees. Proceedings of the Iowa Academy of Science, 60: 707-715

Ruttner F. (1988). Biogeography and taxonomy of honeybees. Springler, New York

Sarwar, M. (2016). Challenges due to bacterial infections of the honey bees and contributions to manage pest problems. International Journal of Entomological Research, v. 1: p.4-10

Steffan-Dewenter I., Potts S.G., Packer L. (2005). Pollinator diversity and crop pollination services are at risk. Trends EcolEvol 20: 651–652

White, G.F. (1920). American foulbrood. Bulletin, US Department of Agriculture, No. 809

Zander, E. (1909) Tierische Parasiten als Krankenheitserreger bei der Biene. Münchener Bienenzeitung 31: 196-204

23

Prospects of Sericulture in Indian Context

Jony Kumar, Mukesh Kumar Mishra and Bhupendra Kumar Singh

Department of Entomology, Banda University of Agriculture and Technology Banda, Uttar Pradesh, India

Introduction

The Greek term "Sericos" is the source of the English word "sericulture". The words "Seri" and "Cos" in this word both refer to culture. Sericulture is the art of rearing the silkworms for the purpose of producing silk. An elastic protein of insect origin, silk is synthesized by the silk gland of the silkworm. Fibrion and sericin, two various types of proteins, constitute the fiber known as silk. Silk is most frequently produced by chins. Following China, India is the second-largest producer of silk. Furthermore, it is the world's largest consumers of silk. The major silk producing state in India is Karnataka. Tasar, Muga, Mulberry, Oak tasar, and Eri silk are all produced in India. One other commercial fabric of animal origin outside wool is silk, which is a gift from nature to humans. Being an eco-friendly, biodegradable, and self-sustaining substance, silk has unique significance in the present day. The promotion of sericulture can have both major financial advantages and an advantageous effect on ecosystem development.

India is the world's fifth-largest producer of silk and practices sericulture, which has been dubbed an employment-oriented industry. All of the industry's segments, including mulberry cultivation, silkworm seed production, silkworm rearing, reeling and weaving of silk, and collection of byproducts and its processing, generate substantial employment and, as consequently, provide a source of income for rural and tribal people. Mulberry sericulture has been a traditional occupation in Karnataka, Tamil Nadu, A.P. and Kashmir; Tasar one, in M.P., Chota Nagpur Division and Orissa; Muga one, in Assam, Nagaland, Tripura; and Eri one, in Assam's and West Bengal. Due to this peculiar nature, the Indian planners have identified sericulture as one of the best-suited occupations for ideal growth and development of rural India.

Sericulture has been strongly promoted in both traditional and nontraditional areas by central and state level government silk departments. With the start of major development activities, it is projected that sericultural activities will begin to pick up speed in the whole country, permitting for the gradually combine of employment opportunities. This may help to alleviate the acute unemployment trouble in rural India and, in turn, to some extent slow down rural-to-urban migration. Sericulture is an agro-based cottage industry that includes interdependent rural, semi-urban, and urban-based activities; thus, in contrast to other agro-based professions, the contribution of women in sericulture industry is dominating, which will be helpful for improving the status of women in family enterprises. In the light of women welfare through Sericulture industry, the Central Silk Board, a statutory organization, carried out a project called "Women Welfare through Sericulture Industry" in the year 2012.

History

An ancient Chinese couple named Lot zu and K wang T are credited with developing silk from silkworms. When a silk cocoon fell into the woman's cup of tea while she was seated beneath the tree, the hot beverage helped the long thread of silk unravel. She discovered a little larva when the silk ran out. It dawned on her that silk comes from a caterpillar larva. This spread when she told everyone about it. Around 2850–2750 B.C. in China, sericulture first appeared. By mistake, a Chinese woman named Xi-Ling created the mysterious silkworm cocoon. The Sericulture Project (Source: www.indiaagronet.com). It is known with certainty where silkworms come from. It is generally believed that the origin is in the Chinese Himalayan area. Additionally, there are viewpoints that attribute the origin of the silkworm to the north-eastern Indian Himalaya. C14-Isotope-Test results provide proof.

Sericulture extension was encouraged at the highest governmental level, and sericulture reached out to small farmers and their education. In a short period of time, silk manufacturing has risen throughout about half of the old imperial China. Between the years 200 B.C. and 300 A.D., Chinese immigrants brought the knowledge of silkworm rearing and mulberry farming to Korea and Japan, With the captures, conquests, and population emigration of Persians and Arabs, the empire continue its expansion. Later in the 9th centuries, the Moors introduced sericulture to Sicily and Spain. In the 15th century, sericulture extended from Spain to northern Italy, which was home to the majority of Europe's major producer- and trading centers. The sericulture afterwards spread to England, Germany, Switzerland, and the Netherlands. Breeding efforts in the USA fell behind those in other nations. During this period, there is labor diversity between the manufacture of cocoons and commercial fabric processing that is focused on commerce. As a result, the primary producing nations are still China, India, Turkmenistan, Uzbekistan, Kirghistan, and recently also Brazil, whereas the main processing nations are the USA, Italy, and France (Jahnke *et al.*, 2001).

Moriculture: Moriculture refers to the cultivation of mulberry plants. There are more than 20 species of mulberries, of which four are widely used:

- ***Morus alba*** - White mulberry is a medium-sized, deciduous tree that grows to a height of 25 to 35 meters. A heavy, spreading crown that is frequently larger than the tree's height is present. The white mulberry tree can grow in a pyramidal style or tend to sag. It has a straight, cylindrical, buttress-free bole that may be up to 1.8 meters in diameter. The bark is vertically fissured, dark greyish-brown in color, and exudes a latex which can be both white or yellowish. The leaves are very different in form, alternating, petiolate, cordate at the base, and light green in colour. They can be simple or complex (3-5 lobed), dentate, palmately veined, coriaceous, and caducous, even on the same tree. Axillary and pendulous inflorescence. The blooms are single, inconspicuous, and greenish in color, resembling, White mulberry is chiefly used to rear silkworm for silk production. Its foliage can be used as a source of fodder for livestock. The leaves and stems can be cooked as a vegetable. The fruits are edible and can be eaten raw or dried and used as a raisin substitutes.
- ***Morus indica*** - Similar to other members of the mulberry family, *Morus indica* is sometimes referred to as a tiny tree or a shrub, seldom growing taller than 10-15 meters (33-49 feet). The plant has little, silky hairs called as on its branches while it is young, but as it ages, these hairs fall off. Light gray-brown describes the branches. The leaves are linked to the tree by petioles and range in size from 4 to 12.5 centimeters (1+12 to 5 inches) in length and 2.5 to 7.5 centimeters (1-3 inches) in width.
- ***Morus Serrata*** - The Himalayan Mulberry is a large, 20-meter-tall deciduous tree. Up to 2-3 m in circumference, with reddish or grey-brown, scaly bark on older trunks and smooth on younger branches, and delicate shoots velvet-hairy. A 2–5 cm long, velvet-haired leaf stalk carries the individual leaves. The leaf blade is oblong to broadly ovate, 5-15 cm long, 3-10 cm wide, 3(-5) nerved from the heart-shaped base, often 3-lobed, edges coarsely serrated bi saw toothed, tip or apices of lobes tapering-with-a-tail, veins hairy underneath; stipules linear-lance shaped, 1.5-2.5 cm long, membranous. Different trees have male and female catkins. Male catkins are 2.5–5 cm length and hairy. Sepals on male flowers are elliptic-oblong and about 1.5 mm long. They are blunt, hairy, and have filaments that are flattened at the base.

- ***Morus latifolia*** - *Morus latifolia* is a medium-sized tree and a very big shrub with a somewhat asymmetrical 8 m tall and broad crown. Strongly bent and twisted branches on which the leaves are joined in a zigzag pattern produce the most striking characteristic. The extremely broad, ovate to heart-shaped, glossy green leaves range in size from 15 cm to 22 cm. The borders have a pointed tip, are hairy on both sides, and are not lobed or dissected. In autumn, the leaf becomes yellow before dropping. "Spirata" only bears female flowers, and no fruits have ever been seen. Frost may be hard on saplings. Plant ideally in a bright, warm, and protected location.

Mulberry cultivation occurs the most frequently by stem cuttings, root grafts, and seeds. From a mature stem, cuttings which are 22 to 23 cm long, 3 to 4 buds thick and long are produced. These may be planted directly in the soil or first in nurseries before being transplanted. Routine pruning is done after the plants have developed and has two benefits: it promotes growth and encourages the sprouting of new branches.

Harvesting of leaves for feeding larva is done in three ways:

- Leaf picking
- Branch cutting
- Top shoot harvesting

Plucking leaves involves plucking each leaf by hand. In the branch-cutting technique, the entire branch and its leaves are chopped and provided to the larvae of the third instar. The tops of the shoots are cut off during top shoot harvesting, and the 4th and 5th instars get them. The agronomic procedures used to cultivate mulberry trees, such as irrigation, fertilizer use, etc., have an impact on the leaf output and quality. Under ideal circumstances, it is thought that each acre of leaves may yield between 20,000 and 25,000 kg annually. In Japan, 600-650 kg of leaves are said to be needed for spring raising and 500-550 kg for fall rearing to produce one box of 20,000 eggs. The amount of leaves needed in India is around 350–400 for 20,000 eggs.

Fig. 1. Moriculture

Types of Silkworms: There are five major kinds of silk of economic significance that are produced by several species of silkworms, which eat a variety of food plants other from mulberry silks.

All of these commercial types of silk are only produced in India.

1. Mulberry silkworm
2. Tasar silkworm
3. Eri silkworm
4. Oak tasar silkworm
5. Muga silkworm

Table 1. Silkworm species and their host

S. No.	Types of silk	S. Name& family	Host
1	Mulberry silkworm (White or Creamy colour)	***Bombyx mori.L*** (Bombycidae)	Mulberry plant (*Morus* Spp.)
2	Tropical tasar (Coperish in colour)	***Antheraea mylitra*** (Saturnidae)	Arjun (*Terminalia arjuna*)
3	Oak tasar (Temperate tasar)	***Antheraea proylie*** (Saturnidae)	*Quercus* Spp.
4	Eri silk (White colour)	***Samia cynthica*** (Saturnidae)	Castor (*Ricinus cummunis*)
5	Muga silk (Golden yellow colour)	***Antheraea assama*** (Saturnidae)	Som (*Bombycina*)

Mulberry Silkworm- This species provides the majority of the world's commercial silk production, and Mulberry silk is frequently mentioned when discussing silk in general. *Bombyx mori* L., a kind of silkworm, consumes the leaves of the mulberry plant over time to produce mulberry silk. Karnataka, Andhra Pradesh, West Bengal, Tamil Nadu, and Jammu and Kashmir are the primary mulberry silk producing states in India, producing collectively 92% of the nation's mulberry raw silk.

The wings of an adult moth are weak and incapable of supporting flight. They are roughly 2.5 cm length, creamy white in color, and completely unmarked. Two to three days after emergence are the adult life span. Emergence occurs followed quickly by mating.

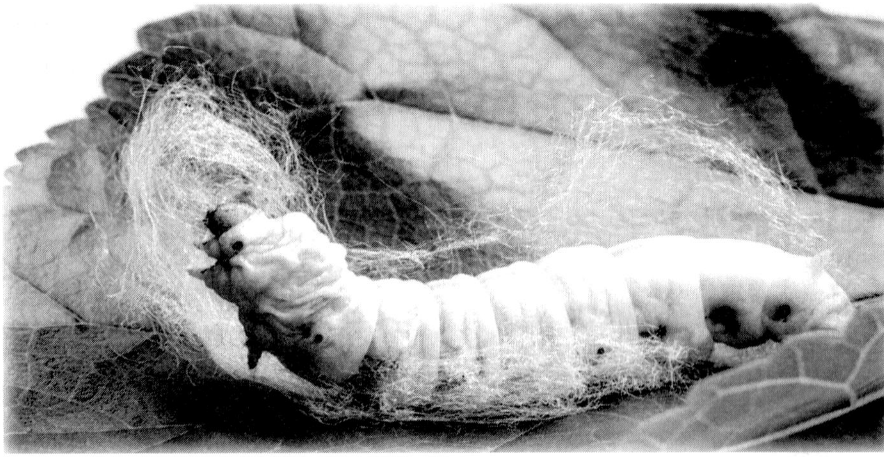

Fig. 2. Mulberry silkworm

Several ***Bombyx mori*** L. larvae's growth and feeding rates

Instars	Weight Increase (mg)	Duration (days)	Leaf Consumed (gm)
1	1 time (0.45)	3-4	17
2	13 times	2-3	80
3	17 times	3-4	320
4	70 times	5-6	2,200
5	10,000 times	7-8	20,000

The number of generations known as "Crops" in the scientific sense of sericulture, when the two instars reach the height of their development and feeding With respect to the moth's race or strain, the generation varies.

- **Univoltine-** Only one generation in a year.
- **Bivoltine-** Two generation in a year.
- **Multivoltine-** If more than two or more generation in a year.

Multivoltine are usually the locally occurring races at same place *e.g.* "Mysore" race producing greenish - yellow cocoons in Karnataka and "Nistari" race producing deep yellow cocoon in West Bengal.

2. Tasar Silkworm

Tasar silk is generated by the silkworm ***Antheraea mylitta*** which mainly thrive on the food plant.

- Sal (*Shorea robusta*)
- Ber (*Zizypus jujuba*)
- Asan (*Termenalia arjuna*)
- Arjun (*Termenelia tomentosa*)

Tasar silkworms are entirely wild and can only be feed outdoors by being attached to food plants, therefore cultivation of the latter may be avoided. The male are smaller and brick-colored, while the female is larger yellowish color and has an obvious eye spot on its wing. Though less glossy than mulberry silk, it has a unique feel and allure. Along with Maharashtra, West Bengal, and Andhra Pradesh, Tasar silk is mostly produced in the states of Jharkhand, Chhattisgarh, and Orissa. Many tribal communities in India rely heavily on their tasar tradition.

Fig. 3 Tasar silkworm

3. Oak Tasar Silkworm

The silkworm, *Antheraea pernyi* J., produces a superior kind of tasar. These originate in south China and are now more thoroughly spread across subtropical and tropical Asia. They consume natural food plants of oak, which are abundant in the sub-Himalayan region of India, which includes the states of Manipur, Himachal Pradesh, Uttar Pradesh, Assam, Meghalaya, and Jammu & Kashmir, unlike the domestic silk moth, which is completely dependent on human care. The majority of oak tasar produced globally comes from China.

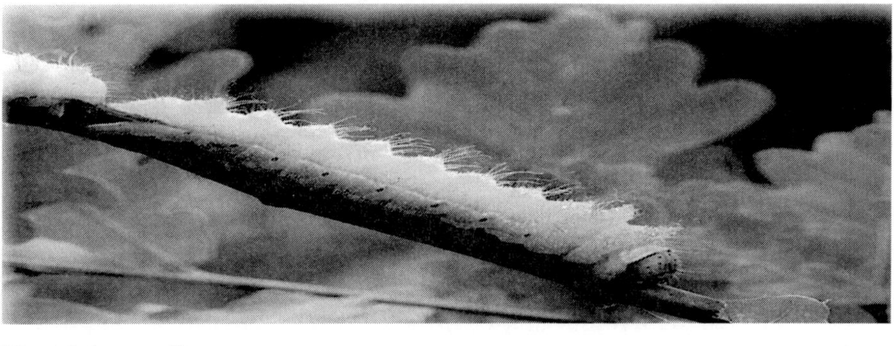

Fig. 4 Oak tasar silkworm

4. Eri Silkworms

The male of Eri silkworm is smaller than the female, and both sexes have wooly white abdomens. Adults are brown in color or have black and green wings with white crescent patterns. Males emerge earlier than females when it comes to adult moths, which happen between morning and midday. The first mating takes place two hours after emergence and continues for until dusk. The males can then be utilized for two to three further matings before being separated. One male is used for mating with the seed crop. These are multivoltine and undergo 5–6 generations every year.

Eri silkworms have a varying longevity of 44 days in the summer and 85 days in the winter. The farmed silkworm *Philosamia ricini,* which mostly consumes castor plants, produces eri silk. Eri culture is a domestic activity done mostly for the protein-rich pupae that the indigenous people eat as a delicacy. These indigenous people utilize the silk creatively to create chaddars (wraps) for personal usage. This culture is mostly prevalent in Assam and the north-eastern states of India. It may also be found in West Bengal, Orissa, and Bihar.

Fig. 5. Eri silkworms

5. Muga Silkworms

Muga is an Assamese term that means brown or amber. Male moths have dark brown and copper-colored wings and bodies, while females have yellowish to brown wings and bodies. Two pairs of wings have eye spots. In addition to color, the male moth may be identified from the female by its somewhat smaller size, thin abdomen, bushy antennae, and sharply pointed forewing tips. The male moth is actively flying during mating with the female. The semi-domesticated multivoltine silkworm *Antheraea assamensis* spawns non-feeding eggs after copulation, which are non-feeding and expire in 7–12 days.

India is the rightful owner of this golden-yellow silk, which is also the state's of Assam's pride. The Muga culture is unique to the state of Assam and is ingrained in that state's heritage and culture. Sarees, mekhakas, and chaddars, among other items, are made from the high-quality muga silk.

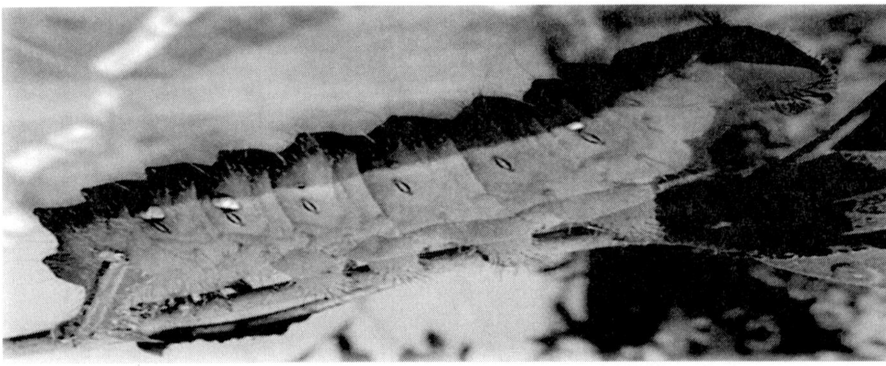

Fig. 6. Muga silkworms

Stages and Production

- The silk moth lays eggs, and after they hatch,
- The larvae eat mulberry leaves.
- The larvae of the silk moth eat mulberry leaves once the eggs have hatched and are ready to hatch. They are 10,000 times heavier than when they were eggs by the time the silkworms are roughly 25 days old.
- At this point, they are prepared to spin a silk cocoon.
- Two glands in the head of the silkworm make silk, which is subsequently pushed out of the body through spinneret-shaped apertures.
- Upon coming into contact with the air, the silk condensed.
- In around two or three days, the silkworm spins about one mile of filament before entirely enclosing itself in a cocoon.
- The amount of usable silk in each cocoon is little due to quality restrictions. Since 1 kilogram of silk requires the production of 5500 silkworms.
- The silk is obtain from the undamaged cocoon by brushing the cocoon to fined the out side end of the filaments.
- The silk filaments are then wound on a reel.

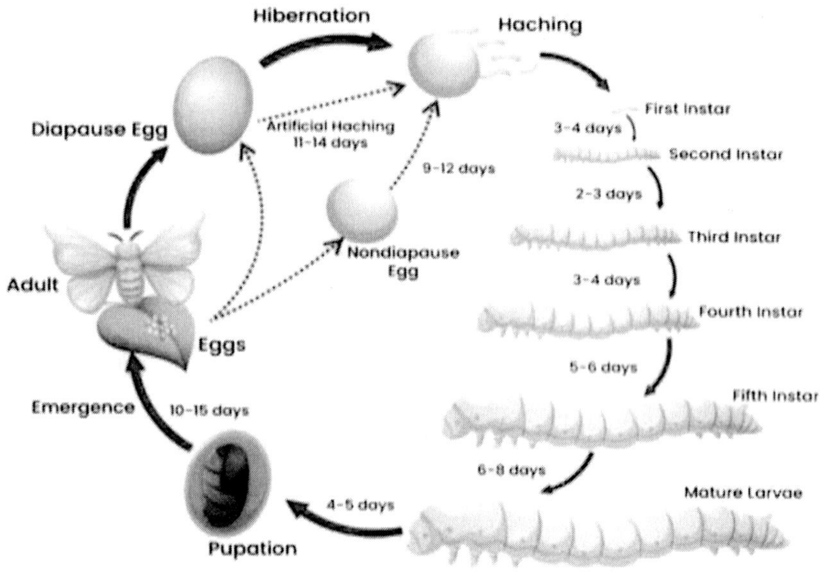

Fig. 7. Life cycle of mulberry silkworms

Eggs: About 300 eggs are laid at a time by the female mulberry silkworm. The eggs have viscous secretions on them that help them adhere to the leaves and papers. The eggs are initially a creamy golden color, changing to a brown hue as they become older and get closer to hatching. The female moth lays eggs and then perishes since she doesn't consume anything afterward. To enable long-term storage, the eggs are kept in a cold environment. Larvae emerge from the eggs at a temperature of 18 to 25 degrees Celsius after two weeks.

Larvae

- The tiny black-colored larvae emerge from the hatched eggs.
- Three to four days after the eggs hatch, the larvae reach their first instar.
- After the first instar, the mouth parts are dark black in color and the larval body is pale black.
- The first instar larvae after two to three days; the second instar larvae after three to four days; and the third instar larvae after four to five days.
- The larvae's body color changes after the second moult, turning completely creamy white.
- The fourth instar larvae appear 3 to 4 days after the third moult and 3 to 4 days after the fourth moult.
- Larvae reach full maturity after their fourth moult, at which point their bodies take on a pale yellowish coloration and turn translucent.

Fig. 8. Fully mature larvae

Pupa

Female pupa

- The female pupa's body is larger than the male pupa's.
- A vertical line can be seen on the ventral side of the eighth abdiminal segment.
- The female mulberry moth's ninth abdominal segment is spotless.
- Male and female cocoons weigh different amounts.

Male pupa

- Male moths lack a vertical line on the ninth abdominal segment and have a thinner body than female pupae.
- On the ninth abdominal segment of the male moth, there is a tiny speck.

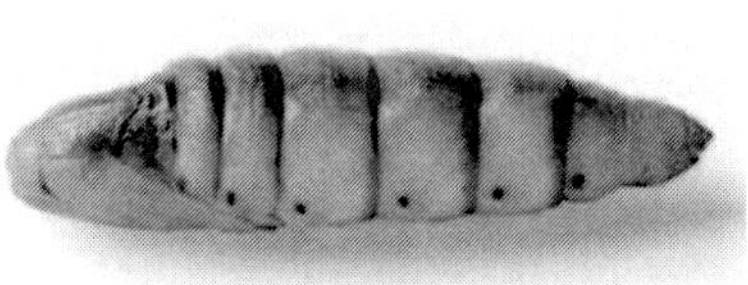

Fig. 9 Adult silkworm moth

Adult

- The mouth parts of both the male and female, which are primarily sucking in nature, are not functioning and neither one can fly.
- They have short, plumose-style antennas and are cremish white in color.
- In contrast to the male, the female has a bigger stomach.
- In contrast to the female moth, the male has much larger antennae.

- In order to attract females for mating, male moths are more active and flap their wings quickly. Within 24 hours of mating, the male moth dies, while the female lays a large number of eggs before also passing away.
- After then, a new stage of the silkworm life cycle starts.

Fig. 10 Adult silkworm moth

What is silk ?

The natural protein fiber known as silk, which comes in several forms that may be woven into textiles, is known as the "Queen of Textiles." Fibroin makes up the majority of the protein fiber that makes up silk, which is produced by specific insect larvae to create cocoons.

Properties

A) Physical property

The triangular cross section and rounded corners of the 5–10μm wide *Bombyx mori* silk fibers are distinctive. Due to a 59-mer repeat sequence of amino acids with slight changes, the fibroin-heavy chain is primarily made up of beta-sheets. Silk has a natural sheen due to the fibrils' flat surfaces' numerous angles of light reflection. Primary filaments (brain), which are bound together by sericin proteins that serve as glue, are spontaneously extruded from two silkworm glands to form silkworm fibers. In contrast to many synthetic fibers, silk has a delicate, smooth texture that is not slippery.

Even though silk is one of the strongest natural fibers, getting it wet can reduce its strength by up to 20%. 11% moisture recovery is a respectable amount. Its elasticity ranges from moderate to poor; even a tiny amount of stretching causes it to remain stretched. If exposed to too much sunshine, it may get weakened. Additionally, especially if kept dirty, insects could assault it.

Due to its weak electrical conductivity, silk is prone to static cling. Silk feels cool to the touch because of its high infrared emissivity.

Silk should either be washed before use in clothing fabrication or dry cleaned since unwashed silk chiffon may shrink up to 8% as a result of a relaxation of the fiber macro structure. Chiffon might still shrink by up to 4% when dry cleaner. Rarely will moderate steaming with a press cloth be able to undo this shrinkage. Both shrinkage caused by molecular-level deformation and progressive shrinkage are essentially nonexistent.

The molecular structure of silk, both natural and synthetic, is likely what causes it to exhibit piezoelectric capabilities in proteins.

The denier, a measurement of fiber linear density, was established using silkworm silk as the reference material. The linear density of silkworm silk is consequently around 1 den, or 1.1 dtex.

Comparison of silk fibers	Linear density (dtex)	Diameter (μm)	Coeff. variation
Moth: *Bombyx mori*	1.17	12.9	24.8%
Spider: *Argiope aurantia*	0.14	3.57	14.8%

Clothing Due to its absorbency, silk is a comfortable material to wear in warm weather and when exercising. Due to its limited conductivity, warm air is kept close to the skin in cold conditions. garments items such shirts, ties, blouses, formal dresses, high-fashion garments, lining, lingerie, pajamas, robes, dress suits, sun dresses, and Eastern folk costumes are frequently made from it. Silk is a great material for utilitarian clothing since it resists numerous biting insects, including horseflies and mosquitoes that would typically pierce clothing.

Silk is frequently used to make a variety of fabrics, including charmeuse, habutai, chiffon, taffeta, crêpe de chine, dupioni, noil, tussah, and shantung.

Furniture-: Many uses in furnishing call for silk because of its alluring sheen and drape. In combination with another fiber, it is used for carpets, bedding, wall hangings, upholstery, wall coverings, and window decorations.

Industry-: Silk is used extensively in industry and commerce, including in parachutes, bicycle tires, comforter filling, and sacks for artillery gunpowder.

Medicine-: Silk can be used as non-absorbable surgical sutures since the exterior sericin layer has been removed through a particular manufacturing procedure. Due to this procedure, specialty silk undergarments, including those used to treat dermatitis, have lately been developed.

Silk has been given new applications and manufacturing methods for everything from disposable cups to medicine delivery devices and holograms.

Rearing

Rearing house-: Since humidity and temperature are particularly sensitive environmental factors, the room or structure where the silkworms are raised must fulfill the requirements. Avoid wetness, air stagnation from exposure to direct sunshine, and maintain optimum humidity and ventilation to obtain the optimal raising house.

Rearing Equipment

Rearing stands- These are the frame stands that have silkworm raising trays on them. Wood was used to make these.

- **Ant wells:** Because ants pose a severe threat to silkworms, the rearing stand's leg is kept submerged in water combined with pesticides on a regular basis.
- **Rearing trays:** Rearing trays are made of wood and occasionally plastic, and their typical size is 150 x 90 cm.
- **Paraffin paper:** To maintain humidity in the rearing trays, thick craft paper sheets coated in paraffin wax are needed.
- **Foam rubber strip:** To maintain humidity, pieces of 2.5x2.5cm foam rubber that have been soaked in water are placed all around the silkworm raising beds.
- **Chopsticks:** Chopsticks are bamboo rods with a tapered end that are used to pick up tender leaf stages while maintaining hygienic handling and avoiding injury.
- **Feathers:** Feathers are used to prevent harm to newly hatched worms by brushing them together.
- **Leaf chamber:** This chamber is used to store mulberry leaves intended for animal feed. The chamber is completely wrapped in gunny sacks cloths that are maintained wet throughout the summer and dry days.
- **Chopping board, knife, and matting:** Since mulberry leaves are provided to the worms in a chopped form in the early stages of the worms, a wood chopping board is utilized to easily collect the chopped leaves.
- **Clearing net:** These nets, which are made of cotton and nylon with mesh sizes appropriate for the various instars, are used to change the rearing beds so that any leftover leaf fragments and trash are filtered out without disturbing the animals.

Fig. 11 Plastic mountage

Fig. 12 Bamboo mountage

Pre cocoon processing

- **Brushing:** The process of transferring the silkworm to rearing trays is called brushing. Suitable time for brushing is about 10.00 am. Eggs at the blue egg stage are kept in black boxes on the days prior to hatching.
- **Preparation of feed bed and feeding:** After brushing, the bed is prepared by collecting the worms and the mulberry leaves together by using a feather. The bed is spread uniformly using chopsticks. The first feeding is given after two hours of brushing.
- **Bed Cleaning:** Periodical removal of left over leaves and worms' excreta may be undertaken and is referred to as bed cleaning. It is necessary for proper growth and proper hygiene.
- **Spacing:** Provision of adequate space is of great importance for vigorous growth of silkworms.
- **Moulting:** A caterpillar's moult is a very important life event. Feeding ceases when the larva settles in for molting. After the last feed, it is advised to scatter some lime powder to stop early ecdysing larvae from starting to feed and preserve uniform growth throughout. The worms are more prone to muscardien infection during the moulting process, although this might be avoided by dusting them with ceresan lime.
- **Mounting:** Mounting is the process of transferring adult 5th instar larvae to mountages. When a larva reaches full maturity, it loses its color, its size, stops feeding, and begins looking for a spot to attach itself so that it can spin a cocoon and pupate. They are taken by the hand and mounted.
- **Harvesting of cocoon:** Cocoons are manually collected. The cocoons are sorted after harvesting. Inside the cocoon, the larva goes through a transformation process to become a pupa. Pupal skin is brittle and prone to rupture in its early stages. Therefore, early harvest may cause pupa damage, which may harm the silk thread. The developing moth may break threads after a late harvest.

Post cocoon processing

- **Stifling:** Stifling describes the method used to destroy pupae inside the cocoon. It is accomplished by exposing the cocoon to hot water, steam, dry heat, sunlight, or fumigation.
- **Reeling:** The process of removing the threads from killed cocoon is called reeling. The cocoons are cooked first in hot water at 95-97^{0}C for 10-15 minutes to soften the adhesion of silk threads among themselves, loosening of the threads to separate freely, and to facilitate the unbinding of silk threads

Sericulture opportunities Sericulture is a cottage industry that contributes significantly to the creation of jobs and the reduction of poverty. It is one of the most lucrative activities in the rural economy. Indigenous technology is easily accessible and reasonably priced. Returns that are timely and consistent. Hand-woven silk is in high demand and is very well-liked. Possibility of developing a huge production unit and organized sector. Silk is weaved into clothing and sarees. The world's largest silk consumer is India.

Infrastructure of Sericulture in India

Union Government

Central Silk Board (CSB) The Central Agency for the Development of Sericulture and Silk Industry is a part of the Ministry of Textile of the Government of India. The CSB's primary activities include maintaining a four-tiered seed production network, assisting the provincial government in implementing a field development program, promoting silk on a national and worldwide scale, and cultivating and putting policy into practice to promote the structured development of the sector. 300 units and 4000 businesses across the nation carry out CSB's activities, while 800 scientists and 2500 technical employees are assigned to specialist areas across a variety of industry specialties. In order to advance the sericulture and silk industries, CSB invests roughly $60 million USD annually.

Provincial Government: All the provincial Government has separate departments dealing with sericulture and silk industry. Their major responsibilities are; implement field development programme, Extension, Marketing, Seed production and incentive investments among the stakeholders. The investment by Governments to the industry is at level of US $ 50 Million per annum.

Private Sector: The development of the industry is greatly aided by the involvement of the private sector, which also includes non-governmental organizations. The export, post-yarn, and seed production sectors all exhibit private investment.

1. Central Sericulture Research and Training Institute, (CSRTI) Mysore (Karnataka).
2. Central Sericulture Research and Training Institute, (CSRTI) W. B.
3. Central Tasar Research and Training Institute, (CTRTI) Ranchi (Jharkhand).
4. Central Silk Technological Research Institute, (CSTRI), Bangalore, (Karnataka).

References

https://www.indiaagronet.com

https://www.indiaagronet.com/Moriculture

Ragumoorti K.N., Srinivasan, M.R., V. Balasubramani, V. and Natarajan, N. (2207). *et al.*, 2007 Principles of applied entomology pp. (35-51) A.E. Publications No. 10 Sundaram Street-1 P.N. Pudur 641041

Devid Vasantharaj B. and Ramamurthy V.V. (2015). Element of Economic Entomology 8th Edition P 45-55 Brillion Publication New Delhi

Srivastava, K.P. (1993). A textbook of applied Entomology 2nd edition Kalyani Pubilisher, New Delhi. II: SO 7P

24

Lac Culture

***Nahida Afreen*[1], *Kartikey Singh*[1*], *Prince Sahu*[1], *Aman Pratap Singh*[2] *and Paritosh Tripathi*[1]**

[1] *Chandra Shekhar Azad University of Agriculture and Tech. Kanpur-208002 Uttar Pradesh, India*

[2]*Acharya Narendra Deva University of Agriculture & Technology, Kumarganj Ayodhya-224229, Uttar Pradesh, India*

Introduction

Lac, a resinous exudation from female scale insects, has been utilized in India since the Vedic period. Its earliest mention can be traced back to the Atharva Veda, where the insect is referred to as 'Laksha,' and its characteristics and behavior are described. In the renowned Indian epic 'Mahabharata,' there is also a mention of a 'Laksha Griha,' a house made of lac, skillfully constructed by the 'Kauravas' with the help of their architect 'Purocha,' with the malicious intent of setting the 'Pandavas,' their formidable enemies, on fire. The use of lac was well-known to the Greeks and Romans since ancient times. The use of lac was well-known to the Greeks and Romans since ancient times. In Asia, the cultivation of lac insects has a rich history, with indications that it dates back as far as 4000 years in China, where its cultivation coincided with the growth of the silk industry. They hardened resin, produced by small lac insects belonging to a bug family, is used in various applications. To obtain 1 kg of lac resin, approximately 300,000 insects lose their lives. From these lac insects, resin, lac dye, and lac wax are derived. Lac cultivation serves as a significant income source for the livelihoods of forest and sub-forest dwellers residing in regions such as Jharkhand, Chhattisgarh, Madhya Pradesh, West Bengal, Maharashtra, Odisha, and certain parts of Uttar Pradesh, Andhra Pradesh, Gujarat, and the NEH (North-Eastern Hill) region. Over time, the uses of these products have evolved and changed. Promotion of lac and its culture can help in eco-system development as well as reasonably high economic returns. It is a source of livelihood of tribal and poor inhabiting forest and sub-forest areas.

Lac Insect Taxonomy

Kerr provided the initial scientific description of the lac insect in 1782, which was published in the Philosophical Transactions of the Royal Society of London (vol. 71, pp. 374-382). The Indian lac insect, Kerria lacca (Kerr) (order: Hemiptera, suborder: Homoptera, family: Lacciferidae), possesses piercing and sucking mouthparts through which it extracts plant sap (Colton, 1984). It feeds on over 400 plant species (Ramani *et al.*, 2008). The commonly chosen host trees for lac cultivation include Palash **(***Butea monosperma***)**, Ber **(***Zyziphus mauritiana***)**, Kusum **(***Schleichera oleosa***)**, and *Flemingia semialata*. Ramani and Sharma (2010) reported *K. lacca* on *Acacia tortilis* and *Calliandra surinamensis*. According to a survey, lac cultivation contributes approximately 28 percent of the total agricultural income on average (Jaiswal *et al.,* 2006).

Distribution

Lac insects flourish and feed on specific species of tropical trees, leading to its distribution across South-East Asian countries. Currently, lac production takes place in India, Myanmar, Thailand, Malaya, Laos, and the Yuan province of China. Among these, India and Thailand are the primary regions involved in lac production, with India holding a dominant position. Lac cultivation was introduced to Thailand from India.The states of Bihar, Jharkhand, West Bengal, Madhya Pradesh, Chhattisgarh, Eastern Maharashtra, and northern Odisha collectively account for over 90% of the Indian lac production. Additionally, there are pockets of lac cultivation found in Andhra Pradesh, Punjab, Rajasthan, Mysore, Gujarat, as well as in the Mirzapur and Sonebhadra districts of Uttar Pradesh.

Life Cycle: The lac insect is a tiny crawling scale insect that pierces its suctorial proboscis into plant tissue to extract juices, grow, and produce resinous lac from its body. This lac ultimately forms a protective covering around the insect in a structure referred to as a "cell," offering defense against predators.

Life cycle of the lac insect

- Life spans approximately six months and includes several stages such as egg, nymph instars, pupa, and adult.
- The insects reproduce using an ovoviviparous method, where the female lays 200-500 eggs that are already fully developed and ready to hatch upon laying.
- Incubation period is a few hours, giving rise to crimson-red first instar nymphs known as "crawlers," which measure about 0.6 x 0.25 mm.
- The emergence of these nymphs is referred to as "swarming" and can last for about five weeks. During this period, the nymphs crawl on branches until they find soft, succulent twigs to settle on.
- Nymphs congregate in groups of 200-300 insects per square inch, where both male and female nymphs feed on tree sap by inserting their suctorial proboscis into the plant tissue.

- After settling for a day or so, the nymphs begin to secrete resin from glands distributed under their cuticle, except around their mouthparts, breathing spiracles, and anus.
- This resin is semi-solid and hardens upon exposure to air, forming a protective cover around the nymphs. Inside these cells, the nymphs molt three times before proceeding to the adult stage.
- After the initial molt, both male and female nymphs undergo degeneration, losing their appendages and eyes. While still inside their cells, the nymphs shed their exoskeleton twice more, reaching maturity as adults. Both male and female larvae reach sexual maturity in approximately eight weeks.

Adult Lac insect (Male)

Only the male lac insect undergoes a complete metamorphosis, transforming into another form; it loses its proboscis and develops antennae, legs, and a single pair of wings. The male lac insect is red in color, measures 1.2 to 1.5 mm in length It resides in a brood cell, somewhat slipper-like, equipped with a round trap door (operculum) from which it emerges. The adult male is winged and walks over females to fertilize them.

Adult Lac insect (Female)

The female brood cell is larger measuring 4-5 mm in length and globular (pear) in shape, firmly attached to the twig. The female retains her mouthparts but fails to develop any wings, eyes, or appendages. As it develops, it becomes a relatively immobile organism with little resemblance to an insect. Essentially, the females primarily function as egg-producing organisms. As the female lac insect grows, her body size increases to accommodate the growing number of eggs she carries. During this period, lac resin is secreted at an accelerated rate, forming a continuous layer that merges into one cohesive body. After a span of fourteen weeks, the female reduces in size, creating space for both light to pass into the cell and the eggs to reside. Around this time, two yellow spots emerge at the rear end of the cell, which gradually enlarge and turn orange. At this stage, the female has oviposited a substantial quantity of eggs within a space known as the 'Ovisac.' The ovisac appears orange due to the presence of a crimson fluid similar to cochineal, called lac dye, which indicates that the eggs will hatch in about a week. Upon hatching, the larvae emerge, and the entire process begins anew.

Once the life cycle completes and the next generation is about to emerge, the resin-encrusted branches are harvested. These branches are carefully scraped, dried, and processed to create various lac products. To initiate the next cycle a portion of the brood lac from the previous crop is retained for the purpose of inoculating new trees

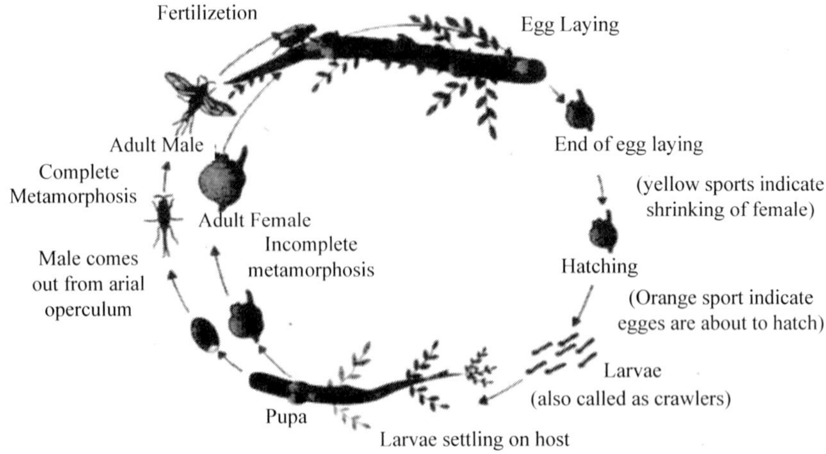

Life Cycle of Lac Insect

Host Plants

Lac insects flourish on specific plant species, where they feed on the plant sap, continuously growing and simultaneously secreting lac resin from their bodies. These particular plants are referred to as host plants. Despite being natural pests to these host plants, lac insects hold a unique position as they are not considered harmful. Several reasons account for this special status:

- They produce a valuable product,
- The host plants are not economically significant, and
- The damage caused by the insects to the host plants is temporary and can be recovered from.

The mentioned varieties of host plants are recognized as lac host plants, with the following being prevalent in India:

- *Butea monosperma* (Palas)
- *Zizyphus spp.* (Ber)
- *Schleichera oleosa* (Kusum)
- *Acacia catechu* (Khair)
- *Acacia arabica* (Babul)
- *Acacia auriculiformis* (Akashmani)
- *Zizyphus xylopyrus* (Khatber)
- *Shorea talura* (commonly known as Sal, grown in Mysore)
- *Cajanus cajan* (also called Pigeon-pea or Arhar)

Among these host plants, Palas, Kusum, Ber, and Khair are of significant importance, while the others have regional significance and are considered less crucial. It is worth noting that the quality of lac directly depends on both the host plant and the strain of lac insects used. Among the different varieties, Kusumi lac stands out as superior in terms of industrial parameters, commanding a higher price in the market. Consequently, the Ber tree is gaining momentum as a potential host for producing Kusumi lac.

Various Strains of Lac

In India, the Lac insect is known to have two distinct strains Kusumi and Rangeeni. The names of the crops are derived from the Hindi months in which they are harvested.

Kusumi Strain

- The Kusumi strain is cultivated on Kusum or other host plants using Kusumi brood
- Kusumi strain, the two crops are as follows:

S. No.	Crop	Inoculation time	Harvesting time	Time period
1.	Jethwi	Jan. / Feb.	June / July	6 months
2.	Aghani	June / July	Jan. / Feb.	6 months

Rangeeni Strain

- The Rangeeni strain thrives on host plants other than Kusum.
- Rangeeni strain yields two crops.
- The lac from the Rangeeni crops is harvested while it is still immature.

S. No.	Crop	Inoculation time	Harvesting time	Time period
1.	Katki	May / June	Oct. / Nov.	4 months
2.	Baisakhi	Oct. / Nov.	May / June	8 months

Among the Rangeeni crops, Aghani and Baisakhi are the main contributors, accounting for about 90% of lac production, while the remaining 10% is contributed by the Kusumi crops. However, the Kusumi crop lac is considered to produce a superior resin due to its lighter color and fetches a better price in the market.

Rearing of Lac insects

The cultivation of lac involves three essential steps:

Host plant selection and cultivation: The first step is to select suitable host plants for lac cultivation. Lac insects thrive on specific trees like kusum, ber, and palas. These trees are chosen based on their ability to provide a favorable environment for the lac insect's growth and development. Once the host plants are selected, they are cultivated and maintained to ensure a healthy and abundant supply of food for the lac insects.

Cultivation of host plant for the lac production in two ways:

1. The plant grows in natural way and the lac-culturist is only protected and care for proper growth of plant.
2. On the other way, a particular piece of land piece of land is taken for purpose and systemic plantation of host plant should be made there. Proper care of plant is required which include

- Regular visit to the field
- Proper Ploughing and Hoeing
- Timely irrigation facilities should be provided
- Suitable application of artificial manure
- Protecting the plants from cattle and human beings

Lac Brood inoculation: In this step, the lac brood, which consists of young lac insects, is introduced to the selected host plants. The brood is usually collected from mature lac colonies and carefully placed on the branches or twigs of the host plants. The young insects then attach themselves to the host plants and start feeding on their sap. This process of introducing the lac brood to the host plants is known as "brood inoculation.

Composition of lac with various constituents and their respective percentages as follows:

- Lac resin: 68%
- Lac wax: 6%
- Lac dye: 1-2%
- Other components: 25%

Processing of lac: It's occurs after a certain period, typically spanning several months, during which the lac insects complete their life cycle, leading to the formation of the resinous substance known as "lac" around their bodies. This lac is harvested from the host plants either in the form of stick lac or by scraping it from the branches. Subsequently, the collected lac undergoes a series of processing steps to eliminate impurities and unwanted materials. The processing methods may involve procedures such as heating, filtration, and shaping, depending on the intended use of the lac. Through this process, different forms of lac, such as shellac or seedlac, can be obtained.

Stick lac: It is also referred to as raw lac or scraped lac, is obtained by carefully removing the encrustation of lac from the twigs of the host plant through scraping.

Shellac: The shellac is the finished product of lac orange to pale yellow in color used across the world. Seed lac is processed into shellac by three methods: Handmade country process, Heat process and Solvent process. It is utilized as a fruit coating for citrus fruits and apples and serves as parting and glazing agents for sweet treats like marzipan, chocolate, etc.

Bleached shellac finds applications in various industries:

- Paint Industry: It is used as a primer for plastic parts and plastic films.
- Aluminum Industry: It serves as a primer for aluminum and aluminum foils.
- Pharmaceutical Companies: Bleached shellac is utilized as a coating for pills, tablets, and gel caps.
- Confectionery: It is employed in the coating of confections, chewing gums, marzipan chocolate, jelly, coffee beans, and more.

By diligently following three key steps - host plant selection and cultivation, lac brood inoculation, and meticulous processing of lac - successful lac cultivation is ensured, ultimately resulting in the production of highly valuable lac resin.

Natural Enemies of Lac Insect and Their Management

Harmful parasitoids like *Aprostocetus* (Tetrastichus) *purpureus*, *Tachardiaephagus tachardiae,* and *Parechthrodryinus clavicornis* have been identified. In addition, 20 predators have been recorded, including significant ones like *Eublemma amabilis* (white moth), *Pseudohypato papulverea* (black moth), and *Chrysopa sp.* (green lacewing). These predators alone contribute to 35–40% of lac crop losses. During different seasons, specific pests like *Chrysopa lacciperda* and *C. madestes* pose a substantial threat to lac insects.

Management

- Various cultural control techniques such as early inoculation (within 10–15 days), trap cropping, and intercropping with plants like okra and cotton can be employed.
- Mechanically, synthetic net container bags with sixty mesh can be used for inoculating brood lac, effectively trapping nymphs while preventing adult predator moths from escaping.
- The application of the biopesticide like *Bacillus thuringiensis* Berliner in the field is an effective method to control *Pseudohypatopa pulverea* and *Eublemma amabilis*.
- Additionally, ant predators like *Solenopsis geminata* and *Camponotus compressus* can be utilized to target the pupae and larvae of these predators.
- Several egg parasitoids, including *Telenomus remus*, *Trichogramm*

atoideabactrae, *Trichogramma brasiliensis*, *T. pretiosum*, and *T. chilonis*, have been employed for managing lac predators.

- The use of systemic insecticides that are safe for lac nymphs and effective against predators can aid in pest control. Combining these various management components holds the potential to reduce pest populations significantly, ultimately leading to an increase in lac crop yield.

Conclusion

It is essential for us to embrace and acknowledge the profound beauty and importance of lac culture. By extending our support to the skilled artisans and communities engaged in this craft, we can truly appreciate its significance. Moreover, we must recognize its invaluable contributions to sustainable development, the promotion of cultural diversity, and the preservation of our collective human heritage. By actively promoting and cherishing lac culture, we ensure that this ancient art form continues to serve as an everlasting wellspring of inspiration, creativity, and cultural pride for generations that follow. Together, let us safeguard and celebrate this timeless tradition, ensuring its continued prosperity and legacy in the ever-changing tapestry of human civilization.

References

Colton H.S. (1984). The anatomy of the female American lac insect Tachardiellalarea. Bulletin Mus. nth. Arizona Flagstaff. 21: 1-24

Jaiswal, A.K., Bhattacharya, A., Kumar, S., Singh, J.P. Evaluation of Bacillus thuringiensis Berliner subsp. kurstaki for management of lepidopteran pests of lac insect. Entomon 2008, 33: 65–69

Kumar, A. Rearing of Lac Insects and Various Useful Products of Lac

Meena, S.C., Sharma, K.K. Species composition and richness of entomo-fauna associated with lac insect, Kerria lacca Kerr in western plains of India. J. Entomol. Zool. Stud. 2018, 6: 684–690

Mishra, Y.D., Kumar, P. Lac culture. In Industrial Entomology; Omkar, Ed.; Springer Nature: Singapore, 2017; pp. 109–155

Mohanasundaram, A., Kaprakkaden, A., Nebapure, S.M., Sharma, K.K., Naaz, N., Mishra, R., Shree, U. Electroantennography and behavioral studies of Eublemma amabilis [Moore] and Pseudohypatopapulverea [Meyr] in relation to volatiles of lac insect (Kerria lacca Kerr.) and its associated products. Int. J. Trop. Insect Sci. 2022, 42: 2313–2324

Ramani R., Sharma K.K., Kumar P. A new record of occurrence of Indian lac insect, Kerria lacca (Kerr) (Coccoidea: Tachardiidae) on Euphorbia pulcherrimaWilld. (Euphorbiaceae). Indian-Journal of- Forestry. Dehra Dun, India. 2008; 31(2): 283-284

Ramani R., Sharma K.K. Record of natural infestation of the Indian lac insect, Kerria lacca (Kerr) (Coccoidea: Tachardiidae) on Acacia tortilis (Forsskal) Hayne and Calliandrasurinamensis Benth. Indian Journal of Forestry. 2010; 33(2): 189-191

Sarvade, S., Panse, R.K., Rajak, S.K., Upadhyay, V.B. Impact of biotic and abiotic factors on lac production and peoples' livelihood improvement in India-An overview. J. Appl. Nat. Sci. 2018, 10: 894–904

Singh, J.P.; Jaiswal, A.K.; Monobrullah, M.D. Green lacewing *(Chrysoperla zastrowiarabica)* incidence on Indian lac insect (Kerria lacca). Indian J. Agric. Sci. 2011, 81: 1087-1089

Singh, J.P.; Jaiswal, A.K.; Monobrullah, M.D. Safety evaluation of some newer pesticides against lac insect (Kerria lacca) for managing predators. Indian J. Agric. Sci. 2011, 81: 465-469

Colour Plates

Chapter 2: Modern Technologies of Insect Trapping

Fig. 4: Insects caught in different layers of mesh of electric trap photograph through webcam (Courtesy: MSSRF, Chennai)

Chapter 5: Role of Bio-Inputs in Pest Management

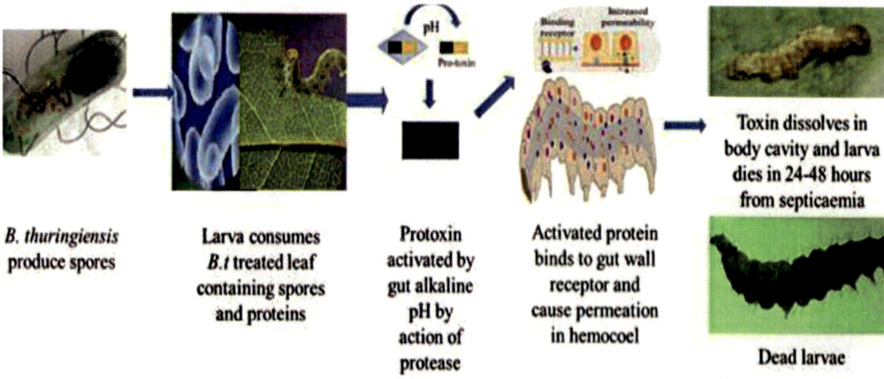

Fig. 1. Mode of action of bacterial pesticide (B. t.)

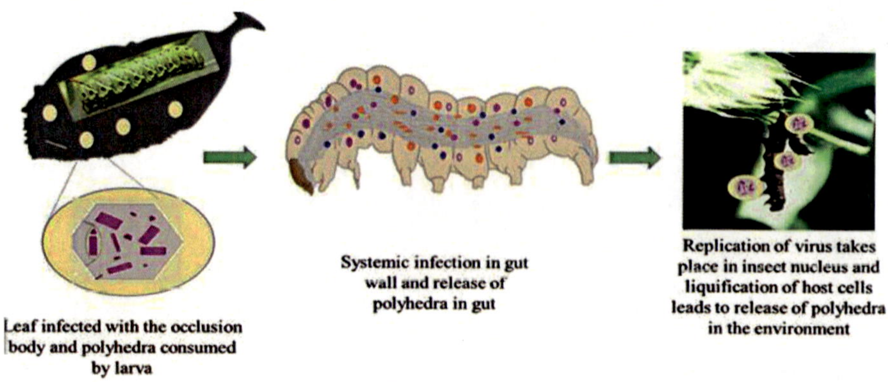

Fig. 2. Mode of action of virus

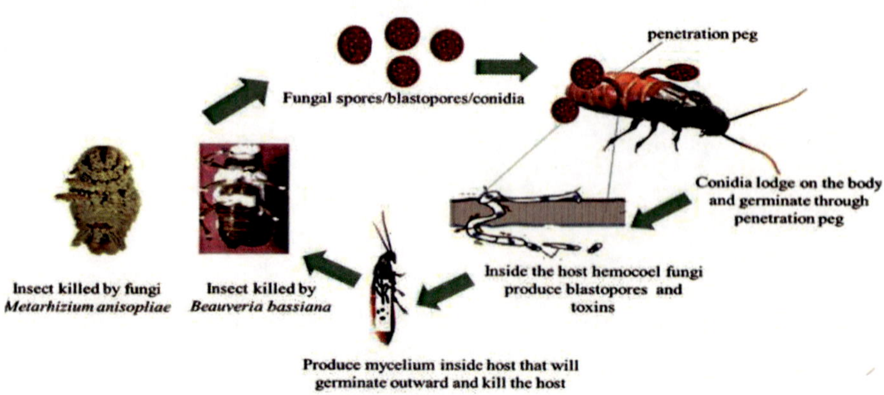

Fig. 3. Mode of action of fungi-based biopesticides

Chapter 6: Biological Control of Insects

T. chilonis Telenomus remus Chelonus blackburni Bracon Sp.

Goniozus nephantidis Campoletis chloridae Brachymeria nephantidis Aphelinus mail

Coccinella septumpunctata

Cryptolaemus montrouzieri

Rodolia cardinalis

Menochilus sexmaculata

Ground Beetle

Platymeris laevicollis

Green lace wings

Cyrtorhinus lividipennis

Preying mantids

Damselflies

Hover flies

Robber flies

Bt infected larvae

Fungi infected larva (Green muscardine)

Fungi infected larva (White muscardine)

NPV infected larva

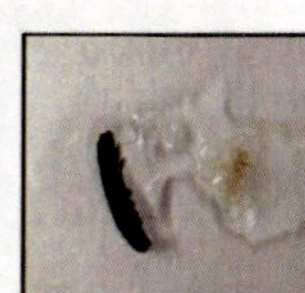

Nematode infected larva

Chapter 7: Role of Entomopathogenic Microorganism in Pest Management

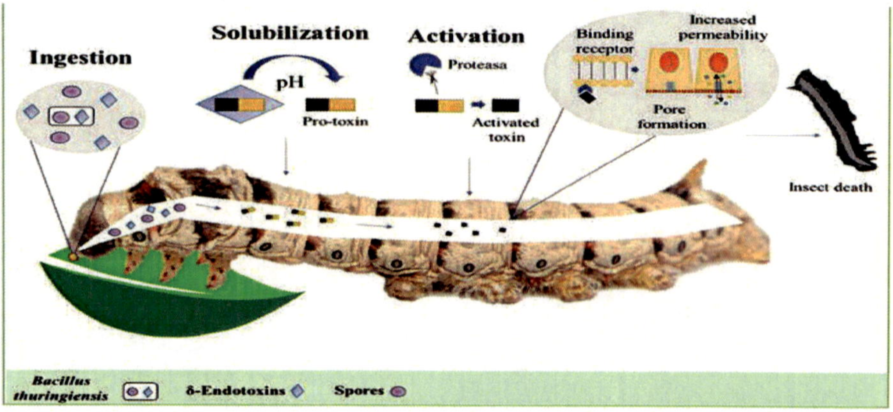

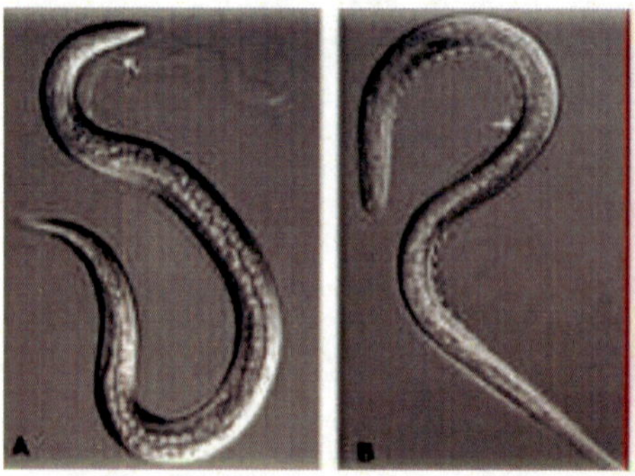

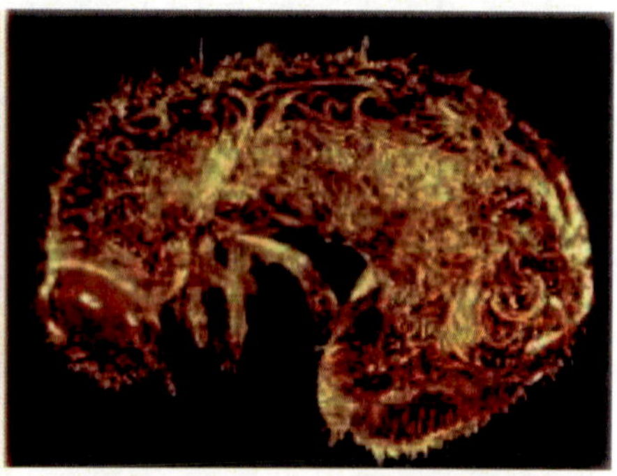

Chapter 11: Chemical Communication Between Plants and Natural Enemies (Carnivorous Insects and Mites)

Fig. 6. Resource-mediated and information-mediated indirect resistance.

Chapter 11: Mass Production Technology of *Trichoderma* spp.

Fig: 1. *Trichoderma* grown on liquid media (PDB Broth)

Fig. 2. *Trichoderma* on solid media (PDA Media)

Chapter 23: Prospects of Sericulture in Indian Context

Fig. 1. Moriculture

Fig. 2. Mulberry silkworm

Fig. 3. Tasar silkworm

Fig. 4 Oak tasar silkworm

Fig. 5. Eri silkworms

Fig. 6. Muga silkworms

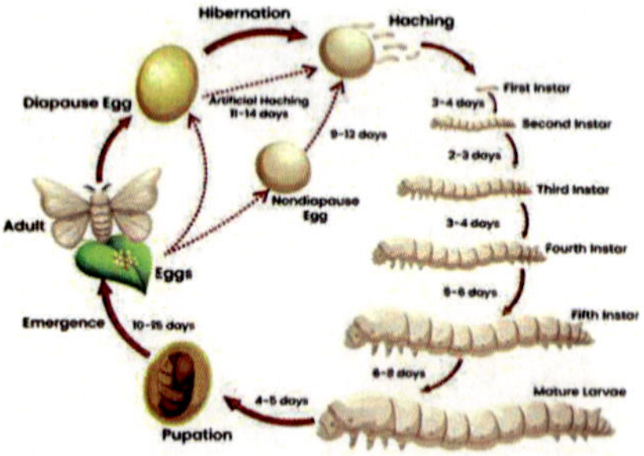

Fig. 7. Life cycle of mulberry silkworms

Fig. 8. Fully mature larvae

Index